SPECTROSCOPIC IDENTIFICATION OF ORGANIC MOLECULES

SPECTROSCOPIC IDENTIFICATION OF ORGANIC MOLECULES

Mohamed Hilmy Elnagdi
Cairo University, Egypt

Kamal Usef Sadek
Ramadan Ahmed Mekheimer
Minia University, Egypt

World Scientific

NEW JERSEY · LONDON · SINGAPORE · BEIJING · SHANGHAI · HONG KONG · TAIPEI · CHENNAI · TOKYO

Published by

World Scientific Publishing Co. Pte. Ltd.

5 Toh Tuck Link, Singapore 596224

USA office: 27 Warren Street, Suite 401-402, Hackensack, NJ 07601

UK office: 57 Shelton Street, Covent Garden, London WC2H 9HE

British Library Cataloguing-in-Publication Data
A catalogue record for this book is available from the British Library.

ISBN 978-981-3271-28-9

For any available supplementary material, please visit
https://www.worldscientific.com/worldscibooks/10.1142/11015#t=suppl

Typeset by Stallion Press
Email: enquiries@stallionpress.com

Acknowledgments

The authors are grateful to Dr. Noha M. Elnagdi from the Department of Organic Chemistry, Faculty of Pharmacy, Modern University for Technology and Information, Cairo, Egypt, for checking the accuracy of spectra presented in Chapters 4 and 5. We would also like to express our gratitude to the AvH Foundation, Germany for financing the activities of Elnagdi and Sadek in contacting experts in the area, especially Prof. H. Meier of Mains University, Germany, who provided several 2D NMR spectral measurements and assisted interpretation. We thank Prof. D. Dopp of Desburg University, Germany for the same. In addition, We are grateful to Prof. Dr. J. Elguero, the President of the Spanish Academy of Science, for providing consultancy for this project.

Contents

Introduction

Both Elnagdi and Sadek graduated in 60s and 70s and so did not receive any spectroscopy courses. This was not unusual as it was during this time that the utility of ^{1}H NMR was just beginning to be appreciated. Although IR spectroscopy and UV were somewhat mature sciences, the apparatuses were scarce in the country (Egypt). In their institution at that time, they had neither mass spectrometers nor X-ray diffraction apparatus. During their M.Sc. and Ph.D. studies, the situation had not changed much. The complete lack of results of any spectroscopic investigation in the M.Sc. and Ph.D. work done by one of us drew critical comments from referees, but the work was subsequently published in journals of good repute. The problems dealt with during that time were rather simple ones and structures could be easily concluded, mostly correctly, based only on elemental analyses. The first experience with spectroscopy for one of us was at the Tokyo Institute of Technology while undertaking a Diploma in applied chemistry sponsored by the Japan commission for UNESCO (1972). Although Elnagdi tried to learn NMR as well at that time, the supervisors thought that there was no need for this. Back home, we gradually felt the need to learn this technique, and so we undertook some self-learning to be able to solve research problems; some of these problems were tricky and there were a few occasions where we did reach wrong conclusions that were not corrected,[1,2] but there were many instances in which we correctly solved many sophisticated research problems. We have, from time to time, received help from our colleagues in the west. In this respect, we have to mention the help of Prof. J. Elguero of Institute de Quimica Medicate who participated in solving some structural problems that we will now mention here, Prof. K. S. Hartaki's (Marburg University) who helped us by gifting 2D NMR texts and Prof. D. Döpp of University of Deusburg who provided, from time to time, high-quality spectra for our compounds when such instruments were not available locally. The push for work given by Prof. Stefen Matlin (UK) through the international organization of chemistry in development by providing facility of NMR should be mentioned as through this foundation we have also obtained good quality spectra that surly opened for us several area in our research. We also have to mention the help with spectra that was provided by Dr. Basil Wakefield of University of Salford. Multiple visits to German institutions supported by the Alexander von Humboldt foundation have undoubtedly helped much in producing data of quality better than what could be obtained locally. We also learned much from our experienced colleagues in Germany. Finally, much was learned during Elnagdi's stay in Kuwait as he enjoyed free access to spectroscopic measurements — a situation we do not enjoy in Egypt as one has to pay from an already meager salary for every spectrum to be obtained. Also, the experience and help provided willingly and swiftly by Prof. H. Meier from Mainz University is gratefully acknowledged. We hope that the next generation, for the good of this country, will not suffer the same situation.

In the following text, we will explain the ways we used to solve research problems in the last forty years based on spectroscopy. Of course, Prof. Mekheimer is a much younger graduate, and so he learned the basics of spectral analyses (IR, ^{1}H and ^{13}C NMR and MS) of organic compounds since he was a young student in the third-year Chemistry Department, at the Faculty of Science. He also learned advanced NMR spectroscopy during his Ph.D. in the Lab. of Prof. Th. Kappe, Institute of Organic Chemistry, Karl-Franzens University, Graz, Austria. During a postdoctoral stay at the University of Connecticut, Storrs, USA, he joined the group of Professor M. Smith and learned how to measure the sample on NMR spectrometer (200 and 400 MHz). Now, he has a wealth of experience in that field and teaches the course of spectroscopic methods in organic chemistry for both undergraduate and postgraduate students at the Chemistry Department, Faculty of Science, Minia University. In this book, we will first summarize the theory beyond the utility of the tool and then discuss our problem. We will start with the easiest, *i.e.* X-ray diffraction, then move to the utility of simple ^{1}H NMR, ^{1}H NMR coupling, ^{13}C NMR, 2D NMR, IR and mass spectroscopy.

Reference

1. M. H. Elnagdi, K. U. Sadek and M. S. Moustafa, *Adv. Heterocycl. Chem. 109*, 241 (**2013**).

1 Solving Research Problems Prior to Use of Spectroscopy

1.1. General Considerations

During a stay in Japan, one of us studied the reactivity of 4-arylhydrazo-nopyrazolidin-3,5-dione (**2**) upon cyanoethylation. Equimolecular amounts of reactants afforded mono cyanoethylation products. The problem was that the researcher hesitated because he was unsure if the product was structure **2** or **3,** but he was more inclined to **3** as hydrolysis did not afford the starting **1** but carboxylic acids **4** instead. However, more confirmatory evidence for this conclusion was required. This could be achieved by synthesis of **3** from the reaction of β-cyanoethylhydrazene (**5**) with diethyl mesoxalic acid arylhydrazones (**6**) (Scheme 1).[1]

Fortunately, this structural assignment was proven to be the correct one as these compounds was reprepared later by chemists at L'Oréal, and their utility for creatinine fiber dyes has been patented.[2]

Another similar problem was solved in 1975 by using the same approach. Thus, cyanoethylation of **7** may in fact afford compounds **8–11**. Structures **8** and **10** were excluded by synthesizing the same product *via* condensing benzoylacetonitrile and β-cyanoethylhydrazene (Scheme 2).[3,4]

The product cyclized on reflux in acetic acid, thus excluding structure **9** as well. In this way, the product could be established as **11** based on its structure. Again, the assignment of **11** to the product was the correct one as this compound has since been resynthesized by Eli-Lilly chemists and

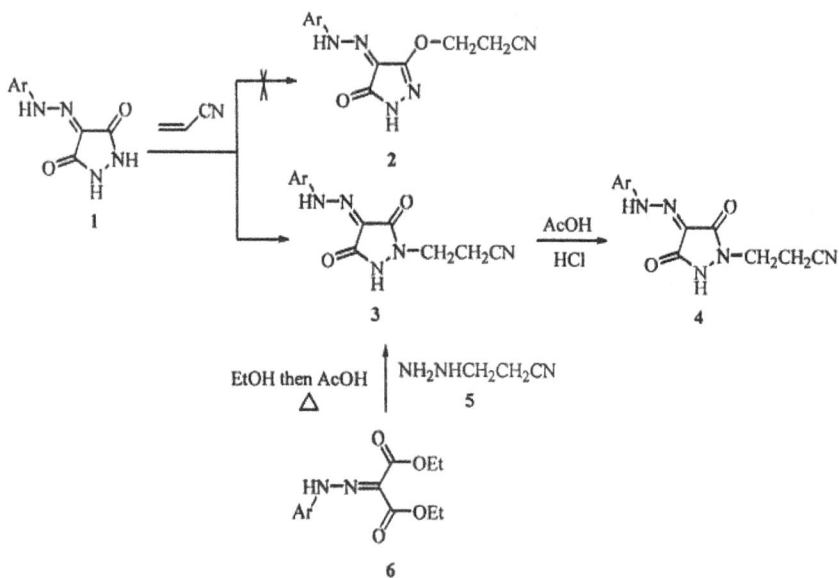

Scheme 1

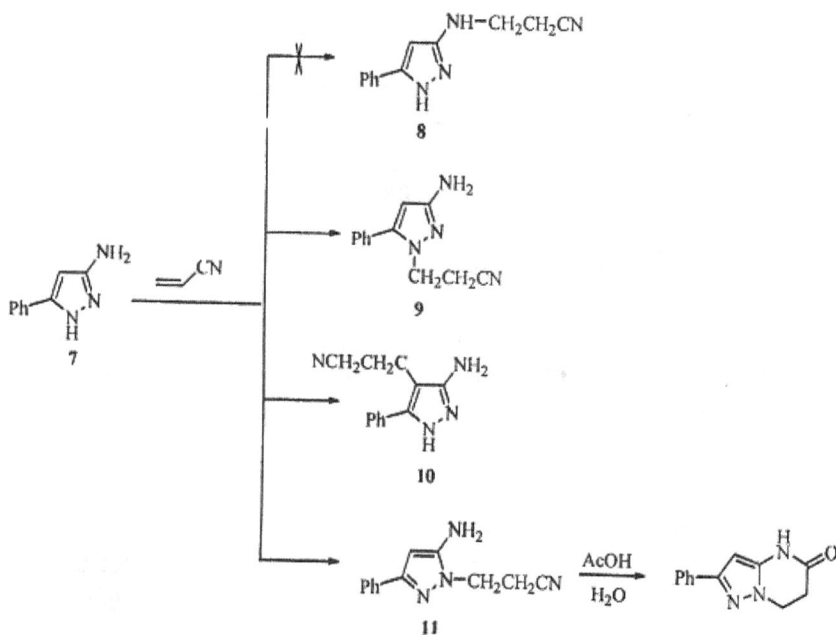

Scheme 2

patented[4] for use. Thus, there is always a way to establish structure without spectroscopy, and structures of many complex molecules, including alkaloids, were established with certainty without utilizing any spectroscopy.[5]

References

1. M. H. Elnagdi and M. Ohta, *Bull. Chem. Soc. Jpn, 16*, 1830 (**1973**).
2. L. Vidal and G. Malle, US patent PCT No. PCT/FR97/00509 (**1997**).
3. M. H. Elnagdi, D. H. Fleita and M. R. H. Elmoghayer, *Tetrahedron, 31*, 63 (**1975**).
4. C. J. Barnelt, R. E. Holmes, L. N. Jungheim, S. K. Sigmund and R. J. Ternary, US Patent PCT No. 1989000418782 (**1990**).
5. J. Mann, R. S. Davidson, J. B. Hobbs, D. V. Banthrope and J. B. Harborne, *Natural Products; The Chemistry and Biological Significance,* Longman (**1994**), Produced through Longman Malaysia WC/01.

2 Utility of X-Ray Diffraction

2.1. General Consideration

This is the most straightforward technique that can help confirm structures in solid state with certainty.

The question is why we did not utilize it prior to 1995?

The answer is simple: measurement on a commercial scale is rather expensive. Moreover, one needs to master the art of crystallization. It was originally assumed that large crystals are necessary, which was later proved to be false. Now, owing to the large-scale technical advancement that has recently occurred and the development of a new generation of instruments, such as those in Kuwait University, it is possible to get X-ray diffraction details for any crystal. However, we were forced to use X-ray while supervising the work of a talented Kuwaiti lady in 1994 (H. Al-Awadi). We were undertaking a simple extension of our thieno[3,4-d] pyridazine chemistry,[1-3] one that is well established with plenty of applications, as assessed by us and others[1-6] (Scheme 1).

The student treated **1** with acryronitrile and obtained a product for which no simple structure could be concluded (Scheme 2). We sent a sample of the resulting powder to Prof. J. Elguero in Madrid. The powder was crystallized by slow evaporation of a saturated acetic acid solution, and X-ray diffraction proved that it was indeed compound **2** (Fig. 1). It was concluded also from X-ray data that the hetero ring adopts a rather flat, distorted boat confirmation owing to the presence of fused rings.

The formation of NH intramolecular hydrogen bonds to carbonyl atom weakened the C=O bond, which had a bond length of 1.220Å. A comment on X-ray was made by crystallographers in Madrid and, together, we came up with a combined paper.[7]

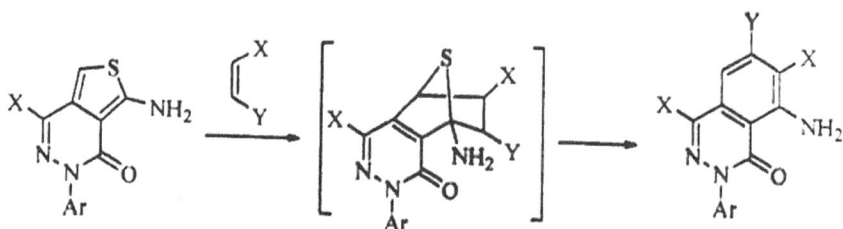

Scheme 1

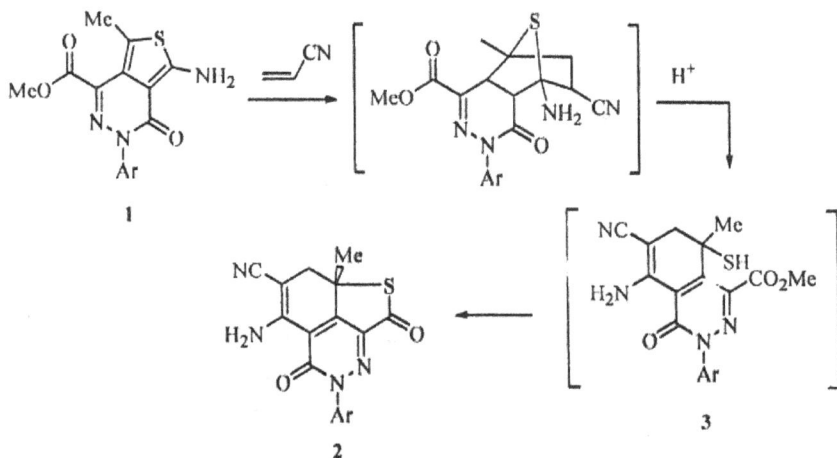

Scheme 2

Another unexpected product that needed X-ray data to confirm the structure was the one obtained on refluxing **4** in AcOH, yielding **5** (Fig. 2).[8]

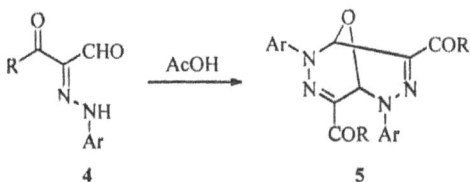

After this, we decided to look back at every structure that we had a doubt in in our previous work and placed emphasis on obtaining X-ray crystal structures for new products for which no decisive structure was reached. Although much success was achieved in this regard thus far, only

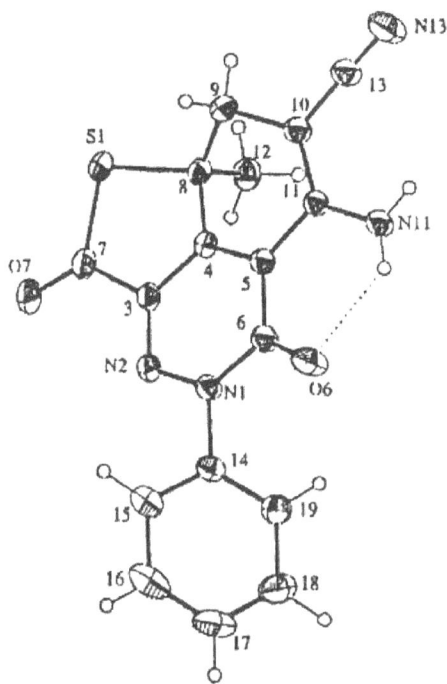

Fig. 1. X-ray structure of compound **2**.

a part of this will be discussed here. For more examples, the reader is referred to our recently published work.

It was believed that compounds with the general formula **6** exist in *syn* hydrogen bonded from although several authors in the 1970s concluded that their spectra is in an *anti* form (**7**).[9-11] X-ray crystal structures obtained in our laboratories[12-15] in fact indicated that this is the preferred geometry for these molecules and although **7** and **8** are conformers, **7** is preferred orientation where the stereo-electronic effect outweighs the stabilizing effect of the hydrogen bonding.[12]

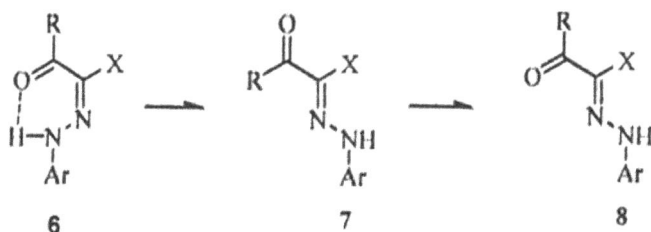

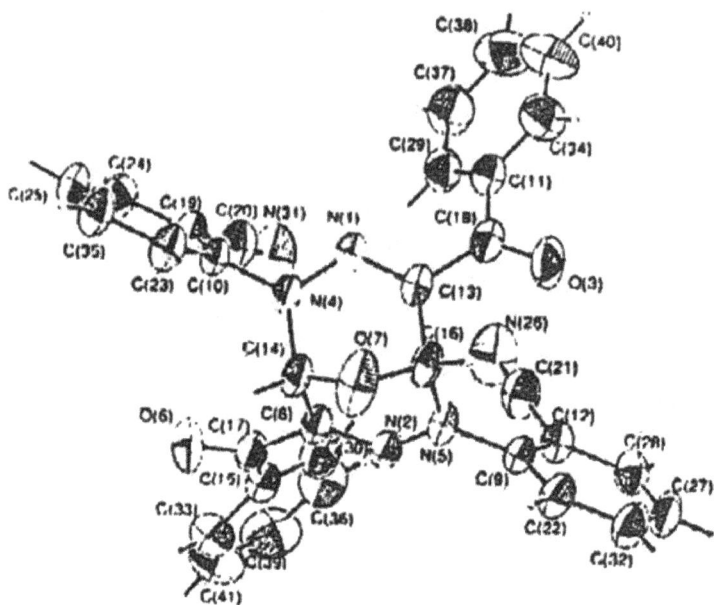

Fig. 2.   X-ray structure of compound **5**.

X-ray crystal structure of compounds **9** and **10** are shown in Figs. 3 and 4.

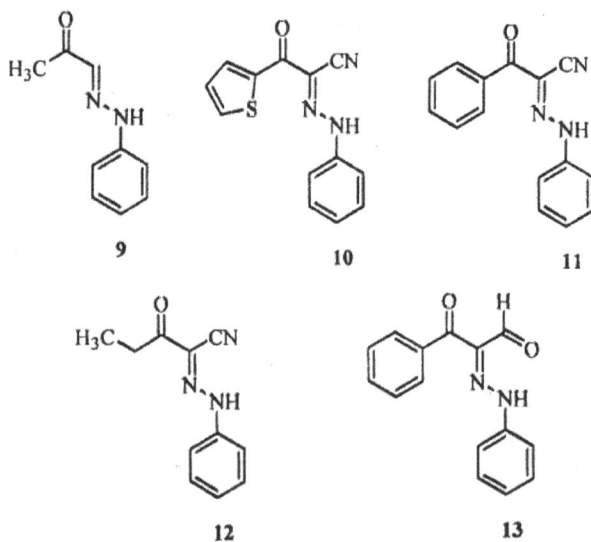

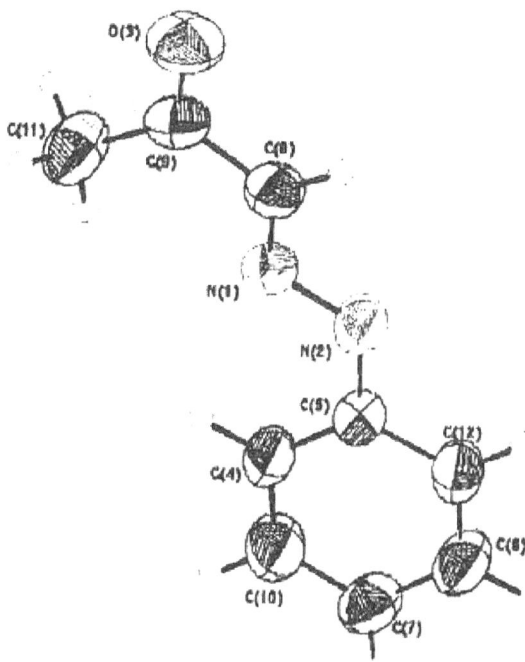

Fig. 3. X-ray structure of compound **9**.

It is interesting to point out that molecule **10** adopts a planar confor-
mation, while in **11** and **13** the phenyl ring rotates almost perpendicular to
plane of the molecule. The planarity of **10** again reveals the role of stereo-
electronic factors. As mentioned earlier, Prof. Hafez and her students
continued analyzing the crystal structures of compounds obtained from
their experiments, and sometimes those obtained from ours, by using
X-rays.[16,17]

Thus, the reaction of **14** with chloroacetone has been suggested to
yield **16** rather than **15**. This was also proven to be correct by X-ray. Also,
in presence of triphenylphosphine, compound **4** reacts with dimethylacetylenedicarboxylate to yield **17** (Scheme 3). We concluded this structure
based on spectral data (will be discussed later), but were still doubtful if
this was indeed correct.[17]

Recently, the data obtained using X-rays (Fig. 5) have clearly revealed
that we did interpret the data correctly.

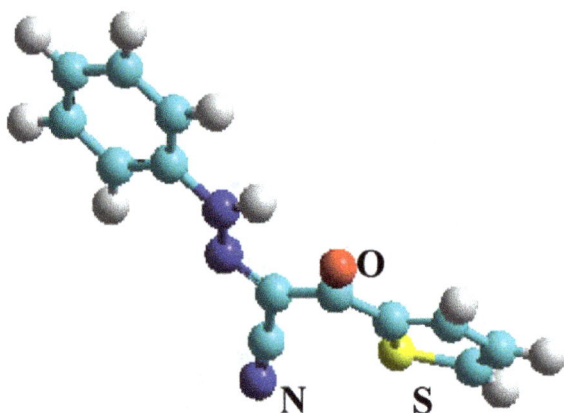

Fig. 4.   X-ray structure of compound **10**.

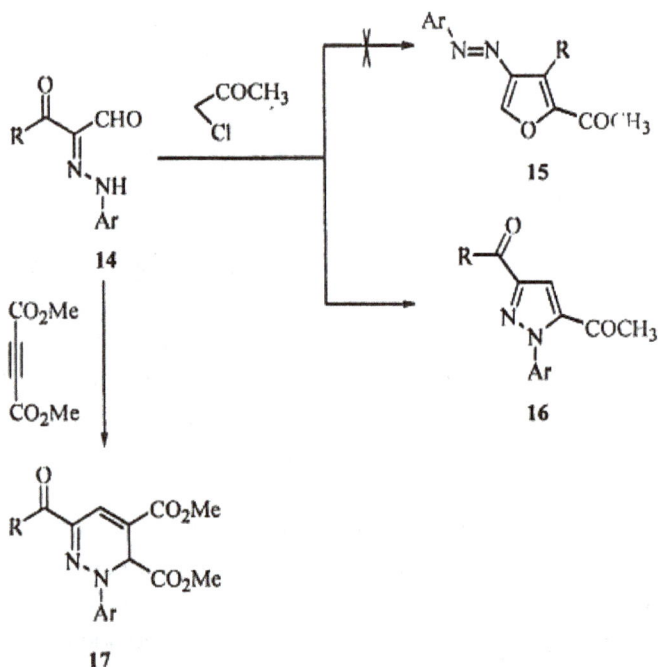

Scheme 3

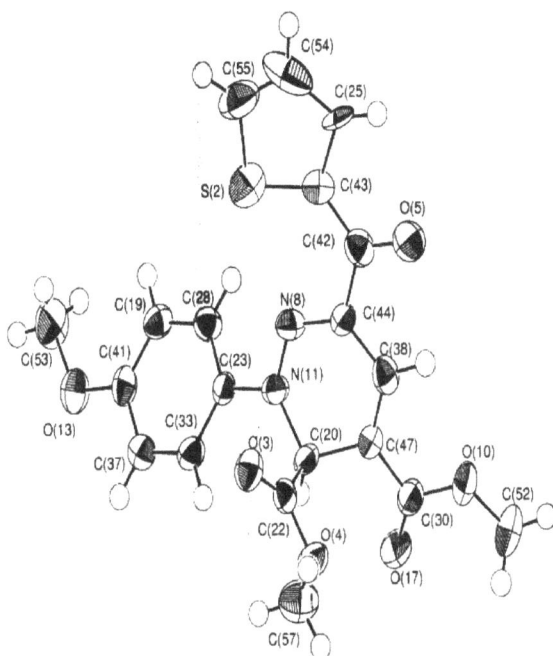

Fig. 5. X-ray structure of compound **17**.

Finally, Dr. Al-Zaydi and her student Al-Shamary suspected that **19** formed from the reaction of **18** with potassium thiocyanate and then underwent a Dimroth type rearrangement and that it was really **20** that was formed, which was proven as being correct by using X-rays (Scheme 4). Thus, they reported a rare rearrangement in thiazole chemistry.[18] X-ray crystal structure determination (Fig. 6) confirmed the structure of a rearranged product.

The success of our group in Saudi Arabia has encouraged us to do the same, thus leaving no doubt with regard to our conclusions, as some of the results appeared, at first sight, quite unexpected. In this respect, Prof. Al-Awadi while performing pyrolysis experiments suggested that the reaction of **14** with hydroxyl amine sulfonic acid gives either 1,2,3-triazoles (**22**), 1,2,4-triazoles (**24**) or isoxazoles (**23**). X-ray crystal structure confirmed that this product was indeed **24** (Fig. 7). The ease of formation of the 1,2,3-triazole system is attributed to the effective delocalization of the N-2 lone pair at the exocyclic carbonyl. This proved to be almost a

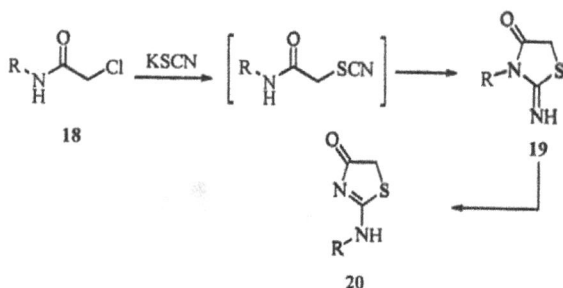

Scheme 4

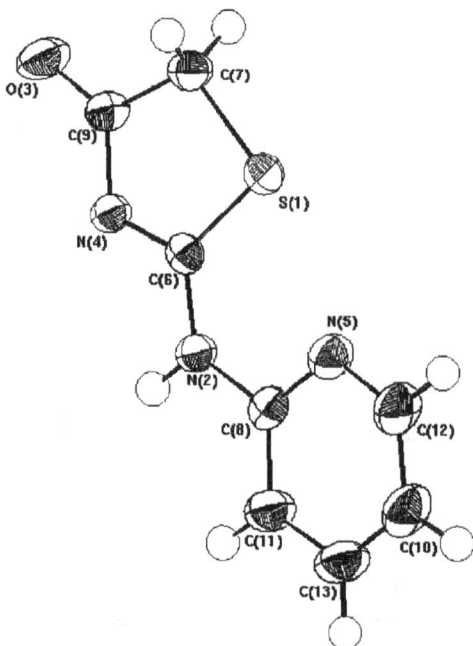

Fig. 6. X-ray structure of compound **20**.

requirement to obtain 1,2,3-triazoles from 2-arylhydrazono-oximes.[19] Thus, in recent work we noted that reacting **14a–e** with hydroxylamine yields amidoximes, which then cyclized readily. The product of cyclization is dependened on the nature of R.[20,21]

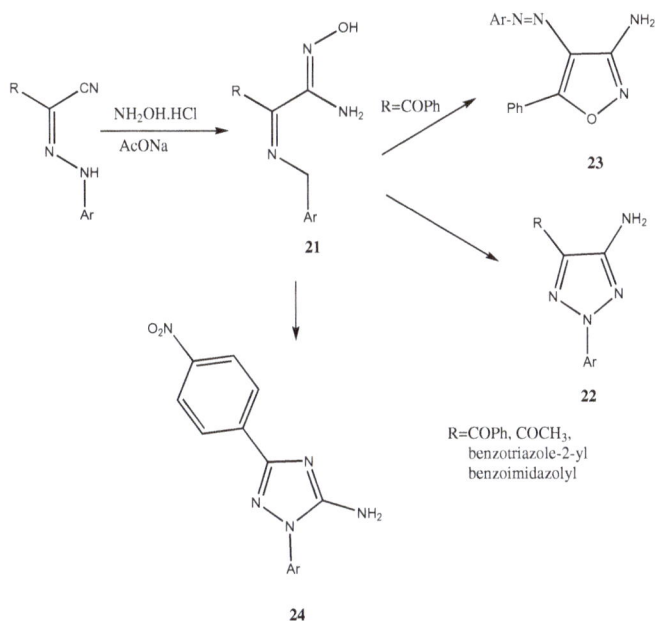

Scheme 5

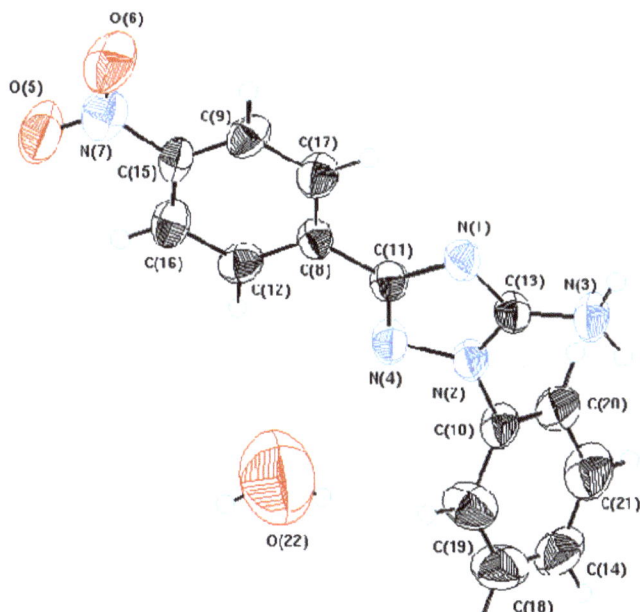

Fig. 7. X-ray structure of compound **22**.

In the case of **24**, the intermediate 1,2,3-tiazoles is rather unstable and thus a rare Beckmann rearrangement occurs (Tiemann rearrangement) yielding **25**. In the chemistry of 2-oxoaryl-hydrazones **26**, we also noted a new Michael addition reaction with **27** and so were inclined to believe that the product formed was the dihydroaminopyridazine **28**. But the referee asked for evidence against **29** (6H pyridazine isomer). Again, using X-rays enabled us to provide the requisite evidence and the paper is now in press (Scheme 6 and Fig. 8).[22,23]

Interlinked with the chemistry of **14**, we then prepared **30** and treated it with ethyl cyanoacetate. To our astonishment, we obtained a product that looked, from the spectral data, to be **31**. As this was quite unusual, we assessed it using X-rays and this proved that our suspicions were correct as the product that was really formed was upon treatment with elemental sulfur in dioxane/pip. under microwave heating was **32**. This was the first X-ray structure of this ring system (Scheme 7 and Fig. 9).[7]

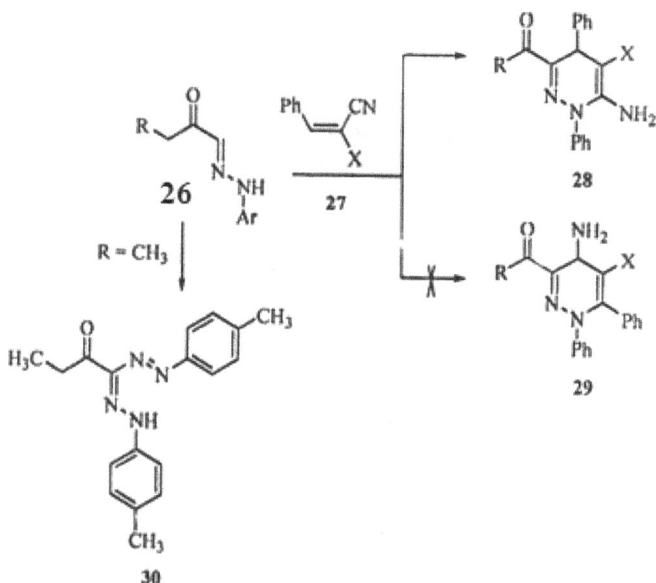

Scheme 6

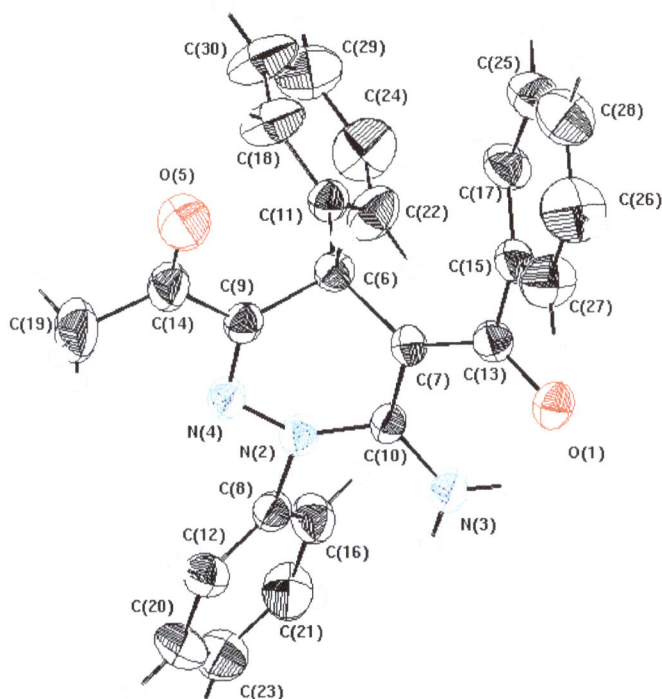

Fig. 8. X-ray structure of compound **28**.

2.2. The Reaction of 1-Arylhydrazonopyruvates with α,β-Unsaturated Nitriles

Upon reacting **33** and **34**, a 1:1 adduct was obtained, and two alternate structures seemed possible (**35** or **36**). X-rays could help solving the X = COPh structure, thus proving that the product formed was **37** (Scheme 8).[24]

2.3. The Reaction of Enaminones with Malononitrile: Novel Rearrangement Reaction Leading to Enedianamide

Reaction of **38** with malononitrile afforded a product due to condensation *via* water elimination. In 1999, we published that the product was

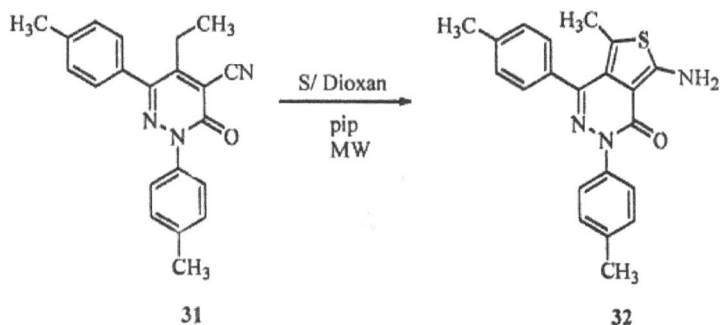

Scheme 7

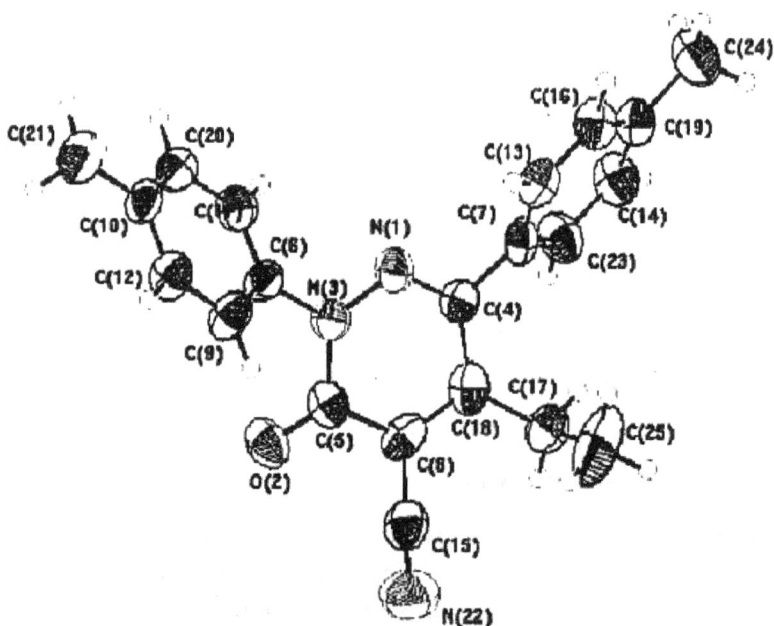

Fig. 9. X-ray structure of compound **31**.

39 but remained uncertain as enaminones should undergo 1,4-addition of nucleophiles.[25] Ten years later, we could obtain the X-ray structure for this product by a reaction with $NaNO_2$ (cf. X-ray CCDC 705064). This is only possible if the product believed to be **39** is really **40**. This

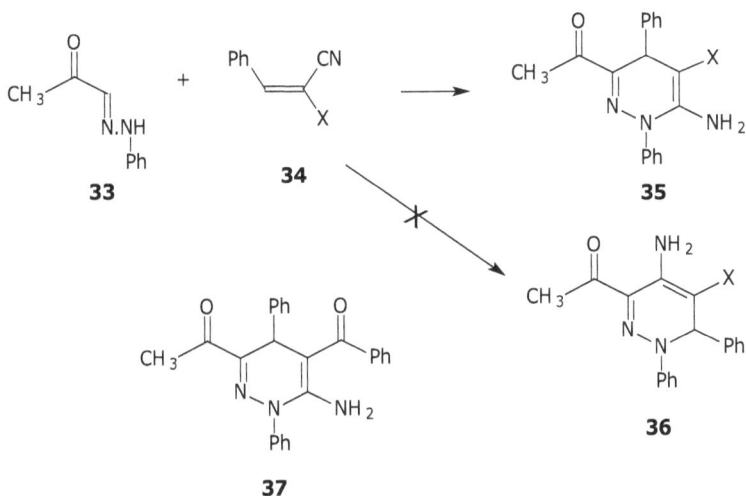

Crystal structure No. CCDC 616675

Scheme 8

was then confirmed via X-ray crystal structure determination to be a derivative of **40**. Moreover, reacting the other derivative further with malononitrile afforded **41** (Scheme 9).[26–28]

2.4. The Reactions of Aminoazoles with α,β-Unsaturated Nitriles

2.4.1. 5-Amino or 7-aminopyrazolo[1,5-a]pyrimidines

The reaction of **44** with **45** has been believed to yield **46**. However, the X-ray crystal structure obtained recently proved that the structure that is formed is really **47** (cf. X-ray No. CCDC 614742, Scheme 10).

2.4.2. 1,2,4-Triazolo[1,5-a]pyrimidines or 1,2,4-Triazolo[3,4-a] pyrimidines

The reaction of **49** with **50** has been proven to yield **51**, not **52**, by X-ray crystal determination.

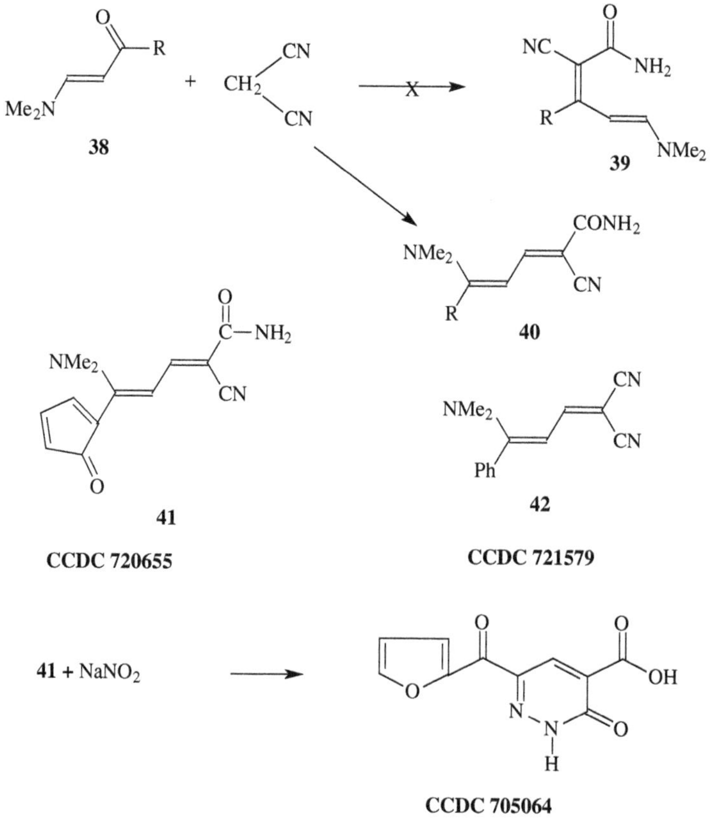

Scheme 9

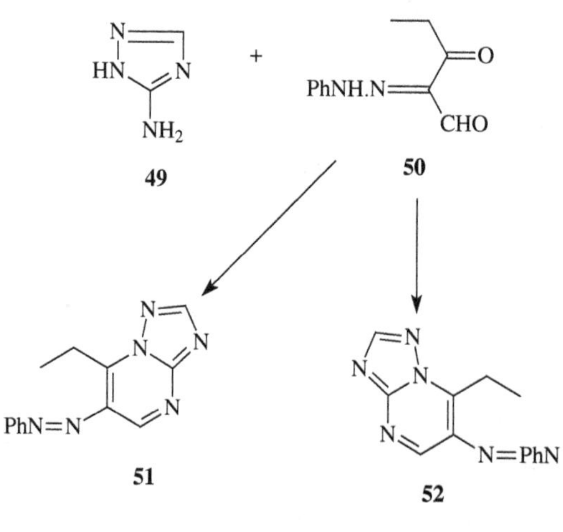

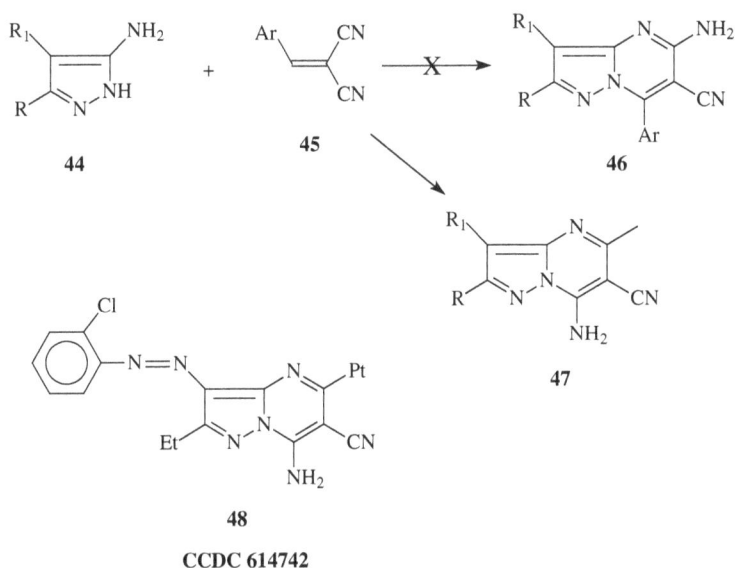

Scheme 10

2.5. The Reaction of Arylhydrazonals with DMAD

This reaction has been proven to yield **53**.

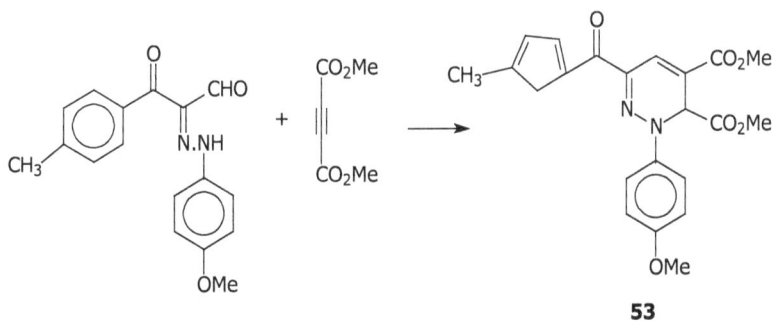

2.6. 1,3,5-Triazolylbenzenes from Enaminols

We reported earlier that a reaction of **54** yields **55**. This conclusion has recently been confirmed by X-ray crystal structure determination of **56**.

54

55

56

CCDC 686291

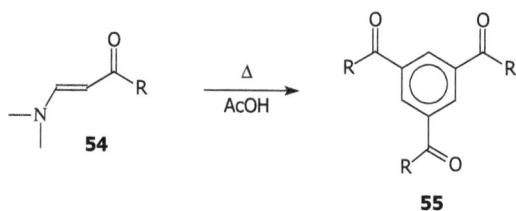

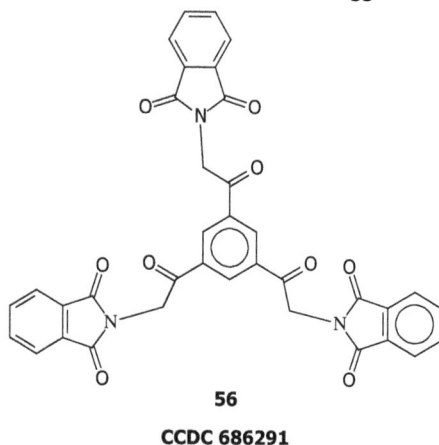

2.7. Unexpected Rearrangements

Recently, we could elucidate the occurrence of several unexpected rearrangements. Thus, we confirmed, using X-rays, that the reaction of **57** with **58** afforded **59**. A mechanism for this remarkable transformation has also been proposed.[29]

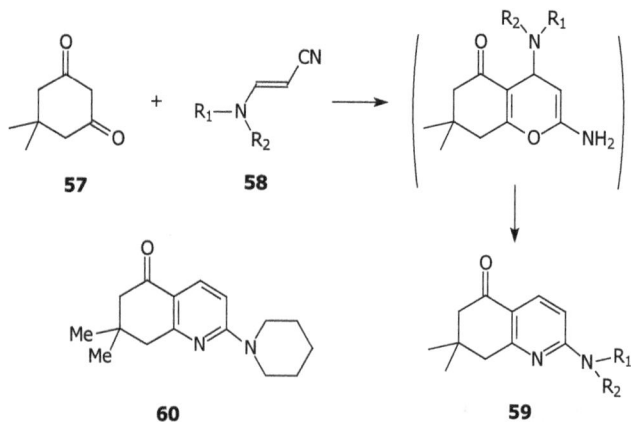

57

58

60

59

Another rearrangement was observed upon reacting **61** (3-(3-(dimeth-ylamino)acryloyl)-4-hydroxy-6-methyl-2*H*-pyran-2-one)with **62** where, instead of expected **63**, compound **64** was formed.[30]

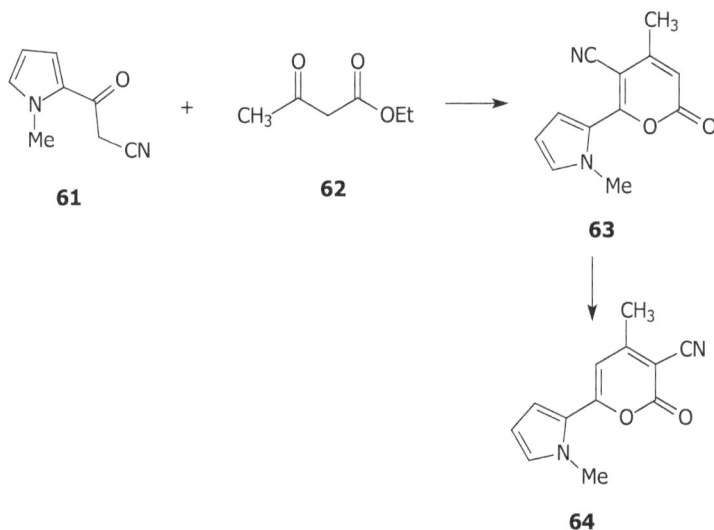

61   **62**   **63**

64

2.7.1. *An interesting story*

An interesting story is the reaction of enaminones **65** with malononitrile in ethanolic piperidine. This product was initially believed to be **66** as spectral and analytical data fitted this structure. But we doubted this con-clusion as, if it was correct, it would be a rare case in which **65** underwent initial 1,2-addition to carbonyl carbon not 1,4-addition, as is the case with other active methylenes (cf. **67** and **68**) that resulted from the reaction with acetyl acetone and ethyl acetoacetate in AcOH/NH$_4$OAc or even formation of **69** upon reaction with hippuric acid. Almost 10 years, later we proposed this in *Molecules* (Fig. 10).

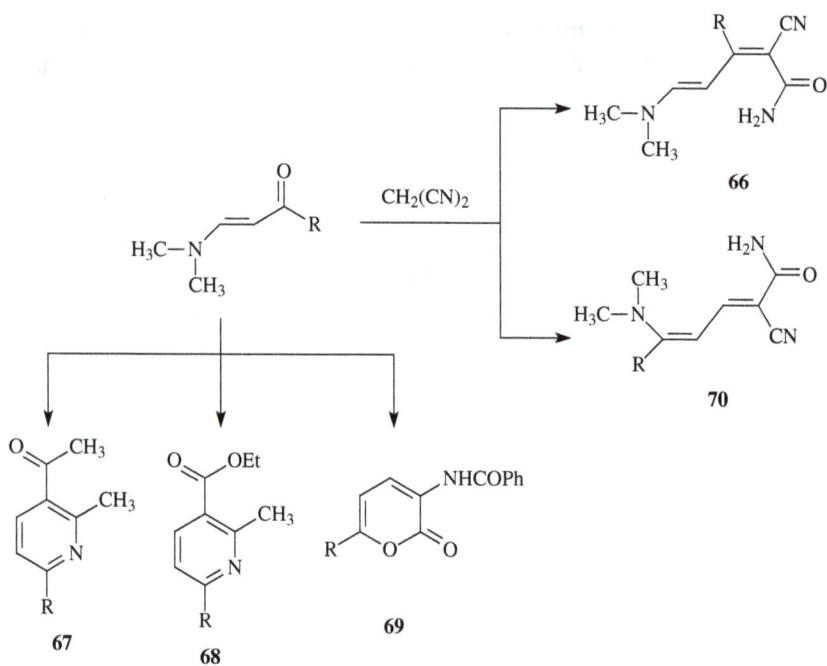

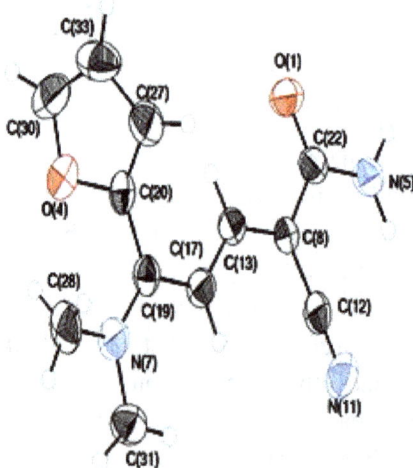

Fig. 10. X-ray for compound **70**.

The reaction of 3-(3-(dimethylamino)acryloyl)-4-hydroxy-6-methyl-2*H*-pyran-2-one **71** with malononitrile under the same reaction conditions revealed that the actual product formed was **75**[20], instead of the previously reported **72**.[21] The structure of **75** was decisively confirmed via X-ray crystallography (Fig. 11).

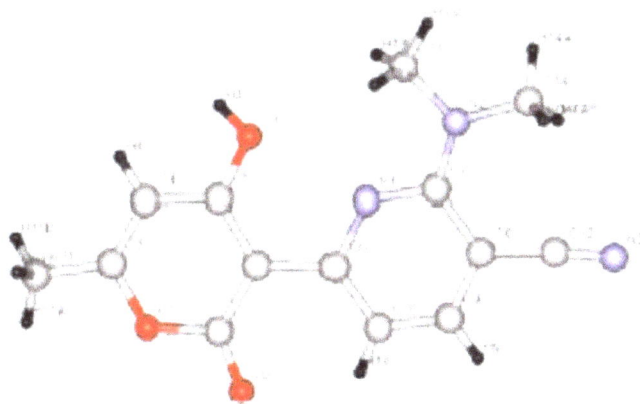

Fig. 11.   X-ray structure of compound **75**.

It is quite strange that a group who are close followers of our work republished this same finding and ignored any reference to our work ten months later although our work was available to them in open access journals even before they initially submitted their article, and surely they had sufficient time to see this when they received their galley proofs.

2.7.2. *Another interesting story*

Yet another interesting story is the result of our re-inspection of the structure of products formed when reacting **76** with ethyl cyanoacetate as it was reported to yield **77**. We suspected this as we believe that pyridazine imines are highly unstable compounds. We noted that [13]C

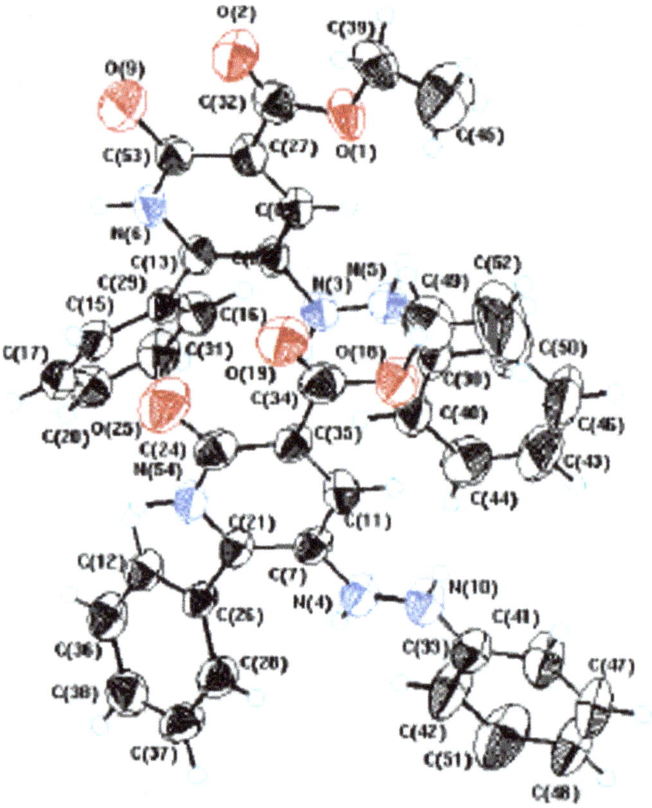

Fig. 12. X-ray structure of compound **73**.

NMR (which was not reported in our published article) has no signal for aryl carbonyl (lowest filed one at δ = 160 ppm). So, X-ray crystal structure could only confirm that the formed product is really **78** — a new arylazonicotinate. We are now conducting several other experiments to further establish this result and hope that upon publishing this preliminary finding, opportunists will not jump on the bandwagon as they have been accustomed to do and ignore our results. In fact, referees of such work should also be blamed as they are assumed to be experts (Fig. 12).

2.7.3. *Yet another very interesting story*

Abdelrazek[31–33] had reported a long time ago that **74** reacted with malononitrile to yield **75**. Again, what was really strange was the 1,2-addition to a carbonyl carbon. We rechecked this finding and noted from spectral data that the product cannot be **75** as ^{1}H NMR and ^{13}C NMR did not reveal signals for sp^3 carbons or protons linked to them as in **75**. We thus have reason to believe that the product is really **76**. However, we could not obtain an X-ray crystal structure for this product. We converted **76** to **77** then to **78** and **79**, and thus X-ray data for this confirming its structure and, consequently, the structure of **76** (Figs. 13 and 14).[34,35]

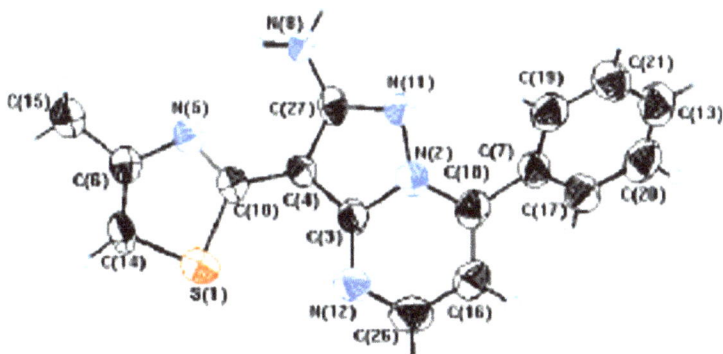

Fig. 13. X-ray structure of compound **78**.

There are several other findings, but what has been explained here is enough to demonstrate the power of X-ray crystal structure determination.

In 2017, Mekheimer *et al.*[36] described the synthesis of novel tetracyclic ring systems benzo[*c*]pyrimido[4,5,6-*ij*][2,7]naphthyridine-1-carbonitriles **13a–c** starting from benzo[*c*][2,7]naphthyridines **83**, which was prepared by reacting 2,4-dichloro-quinoline-3-carbonitrile (**80**) with cyanoacetamide **82** (Fig. 15). Heating **83** with amines **84a–d** at reflux

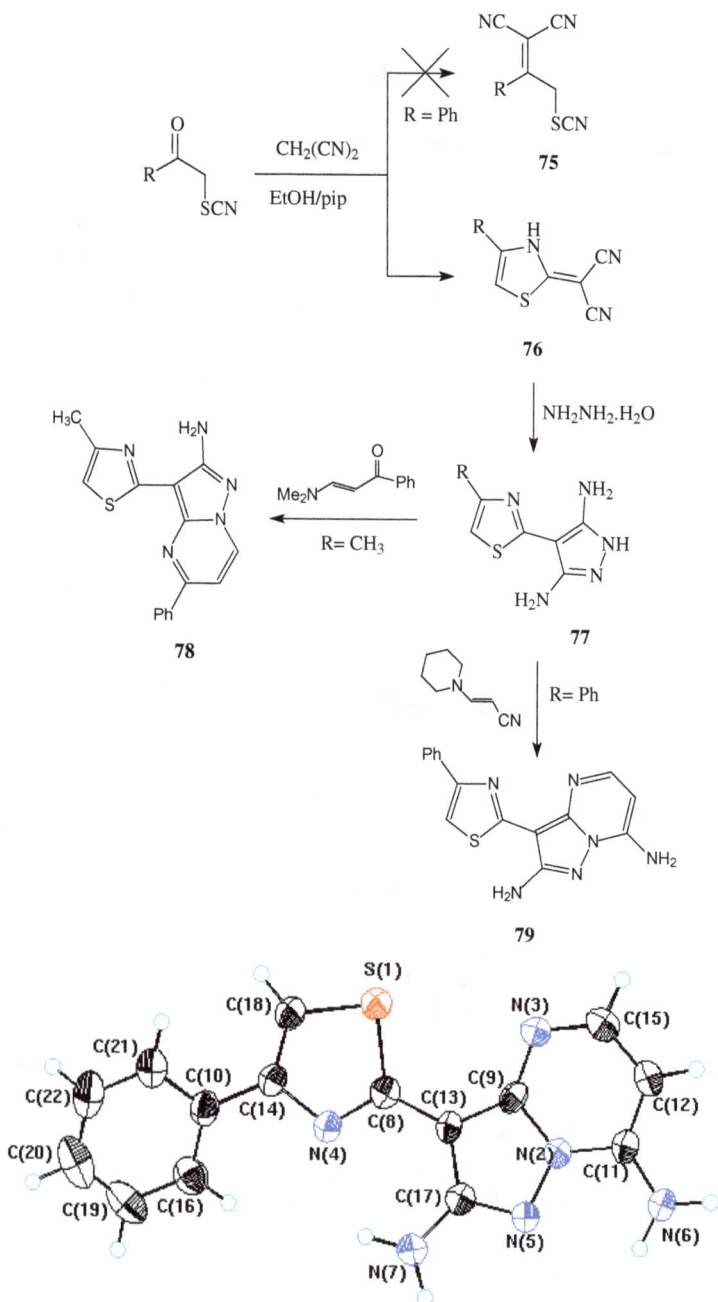

Fig. 14. X-ray structure of compound **79**.

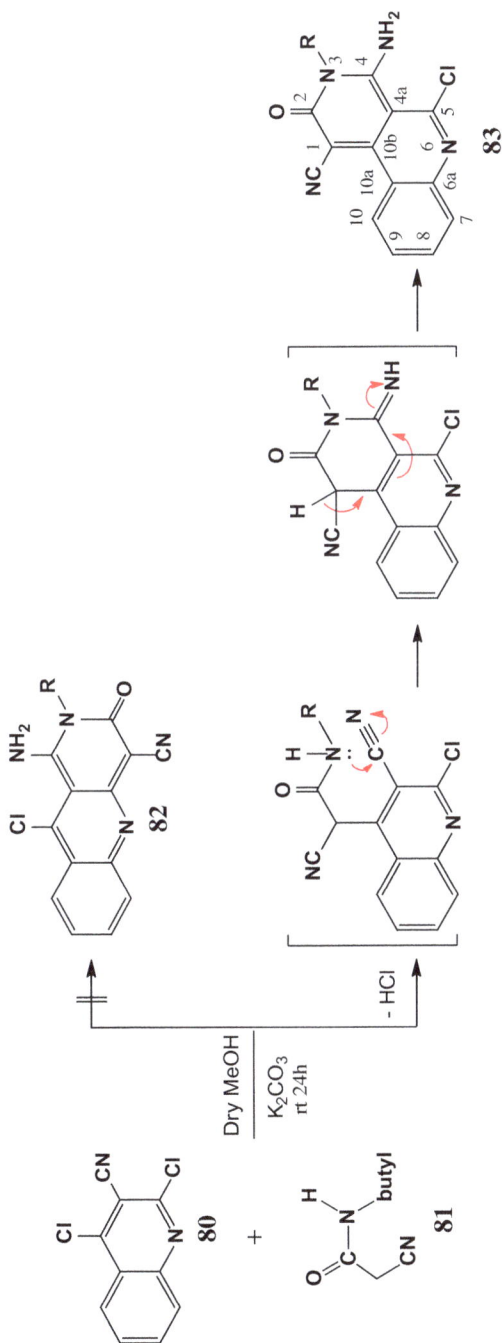

Fig. 15. Synthesis of 4-amino-3-butyl-5-chloro-2-oxo-2,3-dihydrobenzo[c][2,7]naphthy-ridine-1-carbonitrile **4**.

temperature for 2–3 h gave 3-alkyl-5-alkylamino-4-amino-2-oxo-2,3-di-hydro-benzo[c][2,7]-naphthyridine-1-carbonitriles **85a–d** as stable crystalline solids in 74–82% yields. The structures of the products were elucidated using IR, ^{1}H NMR, ^{13}C NMR, mass spectrometry and elemental analyses. Interestingly, the ^{1}H NMR spectra of compounds **85a–d** demonstrated a marked downfield shift for the N–H signals due to an intramolecular N–H⋯⋯N–hydrogen bond interaction.[37] When **85a–c** were heated with acetic anhydride under reflux conditions for 2–3 h, the novel 3,6-dialkyl-5-methyl-2-oxo-3,6-dihydro-2*H*-benzo[c]pyrimido[4,5,6-*ij*]-[2,7]naphthyridine-1-carbonitriles **88a–c** were isolated as the only product (Fig. 16).

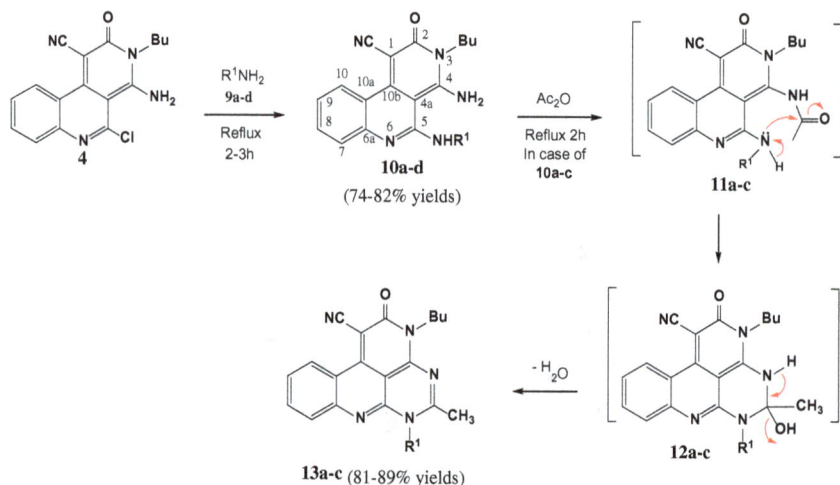

Fig. 16. Synthesis of 3,6-dialkyl-5-methyl-2-oxo-3,6-dihydro-2*H*-benzo[c]pyrimido- [4,5,6-*ij*][2,7]naphthyridine-1-carbonitriles **13a–c**.

The structures of products **13a–c** were established and confirmed based on their elemental and spectral data. Both FT-IR and ^{1}H NMR spectra confirmed the absence of amino group signals (NH, NH$_2$). Single-crystal X-ray

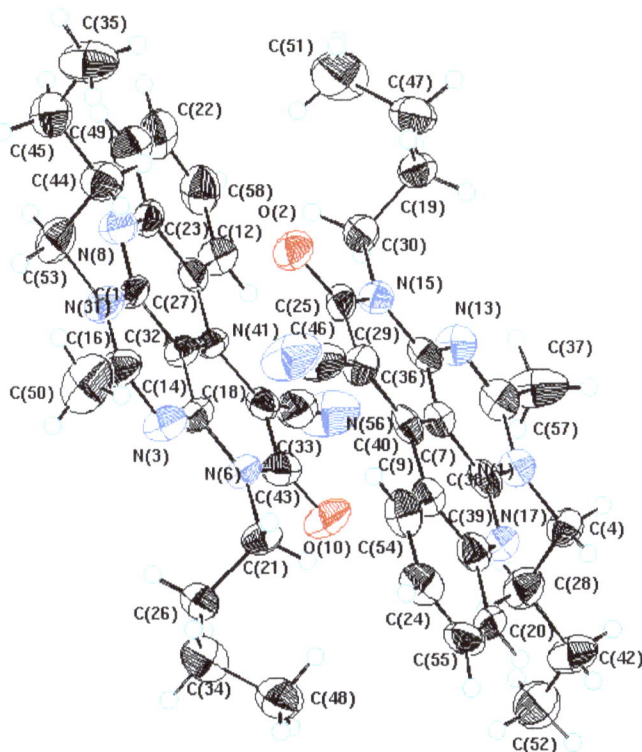

Fig. 17. Single-crystal X-ray diffraction structure of compound **13a**.

analysis of compound **13a** helped to unambiguously confirm the structure of these compounds (Fig. 17).

2.8. Conclusion

X-ray diffraction is definitely a powerful tool for determining the structure of organic molecules in solid state. However, other analytical tools should also be employed to assess structures in the liquid phase. Moreover, the recent tendency to publish only molecule drawings without discussing data like bond length, bond angle and solid-phase resonance form is inconvenient.

References

1. M. H. Elnagdi, A. M. Negm and A. W. Erian, *Lebigs Ann. Chem.* 1255 (**1989**).
2. M. H. Elnagdi and A. W. Erian, *Lebigs Ann. Chem.* 1215 (**1990**).
3. M. H. Elnagdi, A. M. Negm and K. U. Sadek, *Synlett.* 27 (**1994**).
4. F. A. Abu-Shanab, B. J. Wakefielo, F. Al-Omran, M. M. Abdelkhalik and M. H. Elnagdi, *J. Chem. Res.* (s) 488 (**1995**); *ibid* (m) 2924 (**1995**).
5. E. S. Fondjo and D. Döpp, *Arkivoc*, *x*, 90 (**2006**).
6. E. Nyiondi-Bonguen, E. S. Fondjo Z. T. Fomum and D. Döpp, *J. Chem. Soc. Perkin Trans.*, *1*, 2191 (**1994**).
7. H. Al-Awadi, F. Al-Omran, M. H. Elnagdi, L. Infantes, C. Foces-Foces, N. Jagerovic and J. Erguero, *Tetrahedron*, *51*, 12745 (**1995**).
8. Kh. M. Al-zaydi and M. H. Elnagdi, *Z. Naturforsch.*, *59b*, 721 (**2004**).
9. Z. Mostafa, M.Sc. Thesis, Cairo University (**2006**).
10. O. M. E. El-Dusouqui, M. M. Abdelkhalik, N. A. Al-Awadi, H. H. Dib, B. J. Georg and M. H. Elnagdi, *J. Chem. Res.* 291 (**2006**).
11. S. A. S. Ghozlan, I. A. Abdelhamid, and M. H. Ernagdi, *J. Heterocyclic Chem.*, *43*, k (**2006**).
12. M. Kenawi and M. H. Elnagdi, *Spectrochimica Acta Part (A)*, *65A*, 805 (**2006**).
13. S. Aziz, H. F. Anwer, D. H. Fleite and M. H. Elnagdi, *J. Heterocyclic Chem.*, *44*, 877 (**2007**).
14. E. A. Hafez, M. Al-Sheikh and M. H. Elnagdi (unpublished data).
15. M. A. Al-Shiekh, Y. H. Medrassi, M. H. Elnagdi and E. A. Hafez, *Arkivoc*, *xvii*, 36 (**2008**).
16. M. A. Al-Sheikh, M. M. Salaheldin, E. A. Hafez and M. H. Elnagdi, *J. Heterocyclic Chem.*, *41*, 647 (**2004**).
17. S. A. S. Ghozlan, I. A. Abdelhamid and M. H. Elnagdi, *Arkivov*, *xv*, 53 (**2006**).
18. Kh. Al-Zaydi, M. Al-Shamary and M. H. Elnagdi, *J. Chem. Res.*, *6*, 408 (**2006**).
19. H. Al-Matar, S. Ryaid and M. H. Elnagdi, *J. Heterocyclic Chem.*, *44*, 603 (**2007**).
20. S. Almousawi M.S. Moustafa Beil. J.Org.Chem .Beil.J.Org.Chem **2007**.
21. S. Makhaseed, H. Hassaneenand and M. H. Elnagdi, *Z. Naturforsch B* 97, 1, (**2007**).
22. S. M. Al-Mousawi, M. S. Moustafa and M. H. Elnagdi, *Heterocycles*, *75*, 2201 (**2008**).
23. S. I. Aziz, H. F. Anwar, M. A. Al-Apasery and M. H. Elnagdi, *J. Heterocyclic Chem.*, *44*, 877 (**2007**).
24. S. M. Al-Mousawi, M. Al-Apasery and M. H. Elnagdi, *Tetrahedron Lett.*, *50*, 6411 (**2009**).
25. F. Al-Omran, M. M. Abdel-Khalik, M. M. Abou-Elkhair and M. H. Elnagdi, *Synthesis*, 1 (**2009**).
26. A. Alnajjar, M. M. Abdelkhalik, A. Al-Eneze and M. H. Elnagdi, *Molecules*, *14*, 68 (**2009**).

27. S. A. Al-Mousawi, M. S. Moustafa, M. M. Abdelkhalik and M. H. Elnagdi, *Arkivoc*, *xi*, 1 (**2009**).

28. K. Khalil, A. Al-Matar and M. H. Elnagdi, *Eur. J. Chem.*, *1*, 252 (**2012**).

29. M. S. Moustafa, S. M. Al-Mousawi, and M. H. Elnagdi, *Synlett.*, *15*, 2237 (**2011**).

30. H. M. Al-Matar, A. Y. Adam, K. D. Khalil and M. H. Elnagdi, *Arkivoc*, *xi*, (**2012**).

31. F. M. Abdelrazek, A. M. Salah and Z. E. Elbazza, *Arch. Pharm.*, *325*, 301 (**1992**).

32. F. M. Abdelrazek, *Heteroatom. Chem.*, *6*, 211 (**1995**).

33. F. M. Abdelrazek and M. S. Bahbouh, *Phosphorous, Sulfur Silicon Relat. Elem.*, *116*, 235 (**1996**).

34. S. M. Al-Mousawi, M. S. Moustafa and M. H. Elnagdi, *Molecules*, *16*, 3456 (**2011**).

35. S. M. Al-Mousawi, M. S. Moustafa and M. H. Elnagdi, *Arkivoc*, *ii*, 224 (**2010**).

36. R. A. Mekheimer, M. Abdullah Al-Sheikh, H. Y. Medrasi and Gh. A. Bahatheg, *Synthetic Commun.*, *47*, 1052 (**2017**).

37. R. A. Mekheimer, M. Abdullah Al-Sheikh, H. Y. Medrasi and Gh. A. Bahatheg, *Molecular Diversity* (**2017**), DOI: 10.1007/s11030-017-9788-x.

3 Classical NMR as a Tool for Elucidating Structures

3.1. Introduction

As indicated earlier, we lived and worked without ^{1}HNMR equipment in Cairo University till perhaps 1984. We relied basically on elemental analyses and IR spectra, both of which were, at that time, free to use for staff members. There was, however, one event that made it clear to us that further evidence for the suggested structures, especially for polyfunctional molecules, should be provided.

In 1976, we published in the *Indian J. Chem.* that ethyl acetoacetate adds to benzylidene-malononitrile (**1**) to yield a 1:1 adduct.[1] We assumed that the product was structure **2**. A year later, Prof. J. L. Soto,[2] perhaps not having seen our work regarding this, in an unknown journal stated that this adduct was structure **3** based on ^{1}H NMR data (Scheme 1). We read that paper and immediately decided to learn this particular method and apply it by any means possible. This was the beginning; we undertook self-learning, and so, at the beginning, we even misinterpreted some of the data. Of course, now we feel that such mistakes cannot occur. However, we corrected all the new work that we did.

3.2. What is NMR?

It is the study of molecular structure though measurement of interactions between radiofrequency electromagnetic radiation and a group of nuclei immersed in a strong magnetic field. As we will see, an NMR spectrum

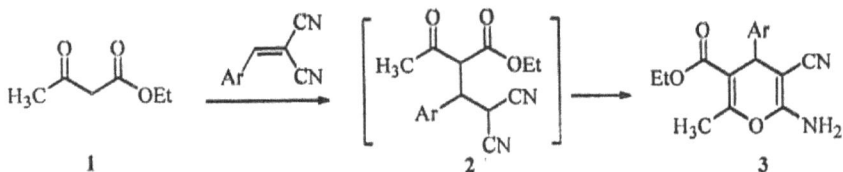

Scheme 1

can provide detailed information that would be difficult, if not impossible, to obtain by any other tool apart from X-ray diffraction.

3.2.1. The story of NMR

It was in 1902 that the physicist P. Zeeman shared the Nobel Prize for discovering that nuclei of certain atoms behave strangely when subjected to strong external magnetic field. It was exactly 50 years later that the physicists F. Bloch and E. Purcell shared the Nobel Prize for putting the so-called nuclear Zeeman Effect to practical use by constructing the first crude NMR spectrometer. During successive years, this instrument has revolutionized chemical and biochemical research. The theory underlying its utility is well illustrated in several texts.[3-5] We will thus give as brief a description as possible on the way we utilized NMR to solve our structural problems.

3.2.2. Why is NMR such a powerful identification tool?

This is because it can treat the nuclei of ^1H, ^{13}C, ^{15}N and P^{31} as functional groups.

Why is this so if the nuclei of an element are the same irrespective of nature of atoms and groups to which such nuclei are attached?

The answer is simple: for the nuclei of certain atoms to be affected by magnetic fields, the field has to penetrate the electron shield around that atom.

The effective charge of that shield depends on the nature of atoms to which the atom is linked. Also, the magnetic field can be enforced or retarded by magnetic fields of neighboring atoms, and both these factors are beyond the power of NMR as a structure elucidation tool.

X	δ_C	δ_H
LiMe	−14.0	−1.94
SiMe	0.0	0.0
Me-H	−2.3	0.23
Me-Me	8.4	0.86
Et-Me	15.4	0.91
Me-NH$_2$	26.2	3.5
FMe	75.2	4.27
SMe	19.3	2.05
Cl	24.9	3.06

Table 1.   Resonance.

3.2.3. Factors affecting chemical shifts

Several factors affect the strength of the magnetic field surrounding the nuclei. To measure these effects, chemists agreed for a zero value for Me linked to Si; this is measured by δ values (ppm). Chemical shifts of protons are the difference between the field resonance of the Me of TMS and resonance of the protons divided by the field power used by the magnet, although this should be a negative value on the T scale, by developing a δ scale which is π-10. But one must to keep in mind that a proton with δ ca 5 ppm resonates at lower field compared to one that resonates at δ ca 2 ppm.

Atoms or groups either shield the methyl they are linked to relative to the SiMe or deshield it. In Table 1, we report such effects using the δ scale.

3.2.3.1. Inductive effects

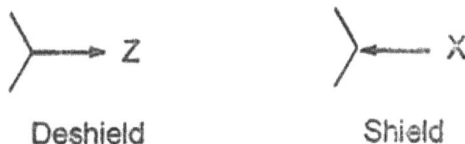

Deshield   Shield

This effect of either shielding or deshielding occurs as a result of electron movement in adjacent multiple bonds.

Table 2. Chemical shifts of C and H in or near multiple bonds.

Compound	δ_C	δ_H
C̲H₃-H	−2.3	0.23
C̲H₃-CH=CH₂	22.4	1.71
C̲H₃-C≡CH2	66.5	1.81
C̲H₂=CH₂	123.3	5.25
C̲H₃-C̲≡C-CH₃	79.2	
C̲H₃-CHO	31.2	2.20
C̲H₃-COCH₃	28.1	2.09
C̲H₃-CN	1.30	1.98
CH₃-C̲HO	199.7	9.80
CH₃-C̲≡N	117.7	

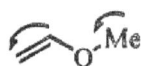

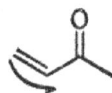

Table 2 gives values for methyl group linked to different multiple bond systems. Generally, double bonds deshield linked carbons and protons, while triple bonds shield.

3.2.3.2. *Ring current*

Ring currents further affect the deshielding of protons outside the ring and shielding of those in ring.

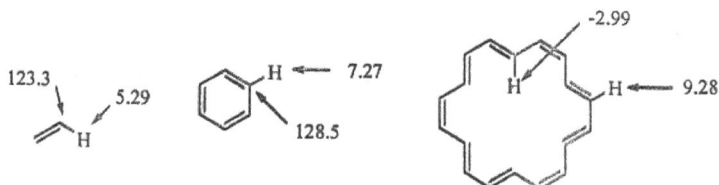

To summarize, we report the general position at which carbons and protons are expected to appear in different systems (cf. Fig. 1).

Now we look at another important feature — this time for protons — multiplicity (coupling) effected by adjacent protons and nuclear Overhauser effects (NOE).

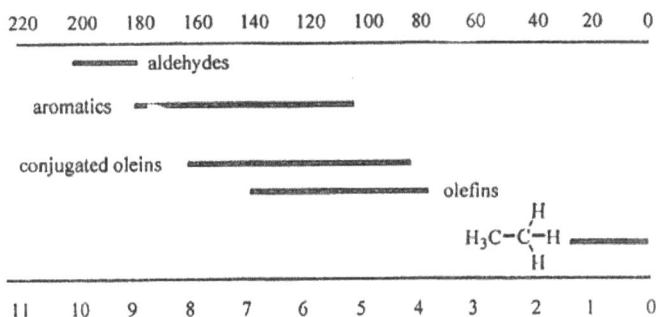

220   200   180   160   140   120   100   80   60   40   20   0

aldehydes

aromatics

conjugated oleins

olefins

$H_3C-\overset{\overset{H}{|}}{\underset{\underset{H}{|}}{C}}-H$

11   10   9   8   7   6   5   4   3   2   1   0

Fig. 1

The effect of magnetic fields on nuclei is either enhanced or opposed by magnets of adjacent atomic protons. This results in the so-called coupling that leads to multiplicity of the proton signal. Protons adjacent to X-atoms will show X+1 signals.

There are two types of coupling as follows.

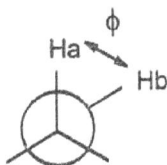

3.2.3.3. *Geminal coupling*

This is always felt unless geminal protons do resonate at exactly the same field at which the irradiated nuclei resonate. The magnitude of this coupling depends on the dihedral angle between protons.

Karplus suggested an empirical equation that relates the magnitude of coupling (J value) with dihedral angle.

Karplus equation

$$3\ Jab = J^0 \cos \Phi - 0.28\ (0\ ^{TM}\ \Phi\ ^{TM}\ 90)$$

$$3\ Jab = J^{180} \cos \Phi - 0.28$$

where J^0 and J^{180} are constants which depend on substituents on carbon atom and Φ is the dihedral angle. This equation leads to the conclusion that J decreases when moving from 0 angle to 90, and when it is 0 then it increases again (Fig. 2).

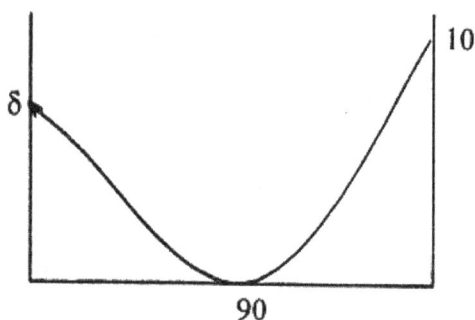

Fig. 2

Table 3. Long-range $^1J_{HH}$ and $^5J_{HH}$ in Hz.

Structure	$^4J_{HH}$	Structure	$^5J_{HH}$
C-CH=CH-CH	0–3		7–8
	1–3		0–1
	0.6–0.9		8–10
HC=C=CH	4–6	=CH-C=C-CH=	0–2
HC≡C-CH=	1–3	HC=C=C-CH=	2–3
	1–2	=CH=C=C-CH=	1–3

The other type of coupling is long-range coupling. Table 3 show some of the values for this.

Having said this, the most important distinctive J values are for the following:

1. Olefinic protons

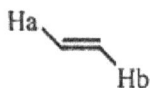

$Ja,b = 12\text{-}18Hz$

$Ja,b = 7\text{-}11Hz$

2. Aromatic and hetero-aromatic protons

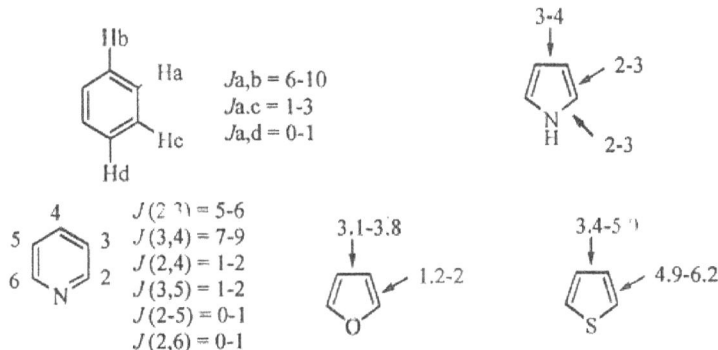

Hb
Ha
Ja,b = 6-10
Ja.c = 1-3
Ja,d = 0-1
Hc
Hd

3-4
2-3
2-3
N
H

4
5
3
6
2
N

$J(2,3) = 5-6$
$J(3,4) = 7-9$
$J(2,4) = 1-2$
$J(3,5) = 1-2$
$J(2-5) = 0-1$
$J(2,6) = 0-1$

3.1-3.8
1.2-2
O

3.4-5
4.9-6.2
S

3.3. Nuclear Overhauser Effect (NOE) and NOE Difference Experiments

To put it simply, protons that are especially proximal enhance each other if the distance between them in space does not exceed 2–4 Å. This phenomenon has been recently utilized for elucidating structures in our laboratories, and we will show our NOE difference curves. (The data presented here is a summary of what is given in other texts which we refer the reader to for further information regarding **3**.)[6]

3.4. Using Classical NMR for Solving Structural Problems in our Research

3.4.1. *Coupling constant and regio-orientation of cycloaddition*

In 1988, we started out in a new direction in our research for which we utilized the pyridazinones **5**, which were prepared earlier by our group[6–8] from **4** for the synthesis of phthlazines and cinnolines **9**. Initially, we added arylidenemalononitriles **1** to **5** and obtained **6**. Repeating this with **7** gave products that were formulated as cinnolines **9**.

Although ^{1}HNMR has been reported, it only provided support for the proposed structures but was not been utilized for discriminating the alternate structure, **8** (Scheme 2).

We then converted **5** into **10** by reaction with elemental sulfur. We reacted the formed thienopyridazine with acrylonitrile. The reaction

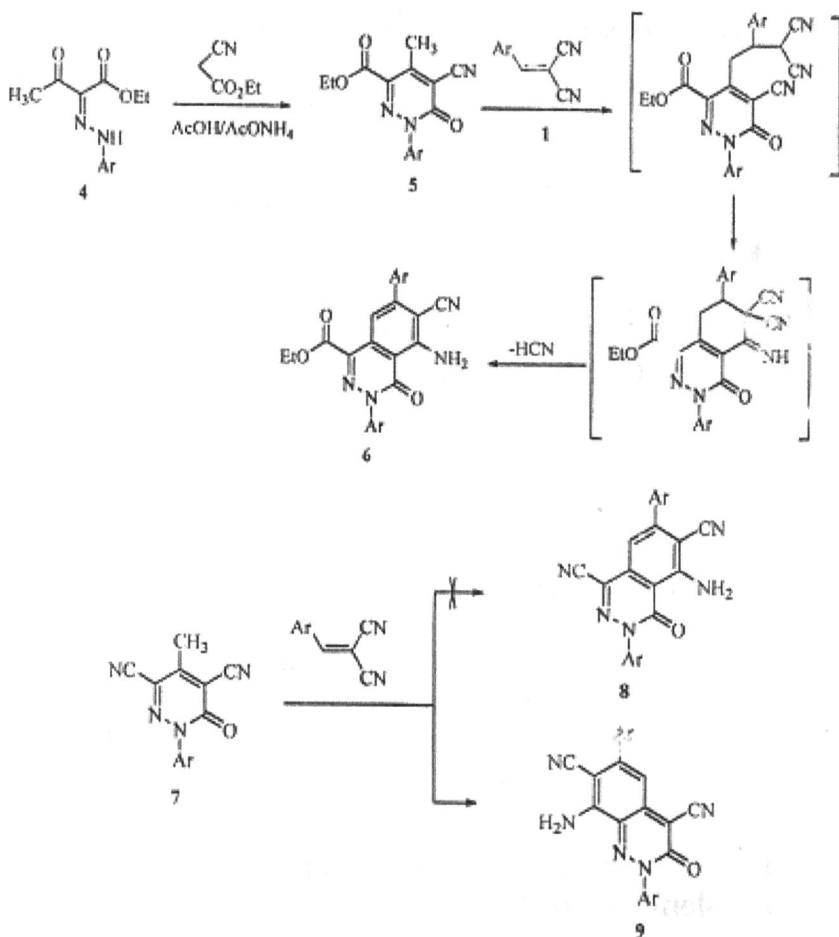

Scheme 2

proceeded readily with great ease. Two regioisomers seemed possible (cf. **11** and **12**). Here, we anticipated that J values for aromatic protons in **11** should be larger than that in **12**, and this proved to be the case[10] (cf. Figs. 3 and Scheme 3).

We have synthesized a variety of derivatives of **13** and noted that all of them react with elemental sulfur to yield compound **14**; typical data for **14** (X = R = H) are shown in Figs. 4 and 5 (Scheme 4).

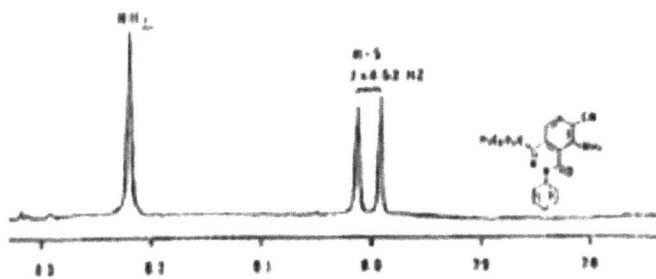

Fig. 3

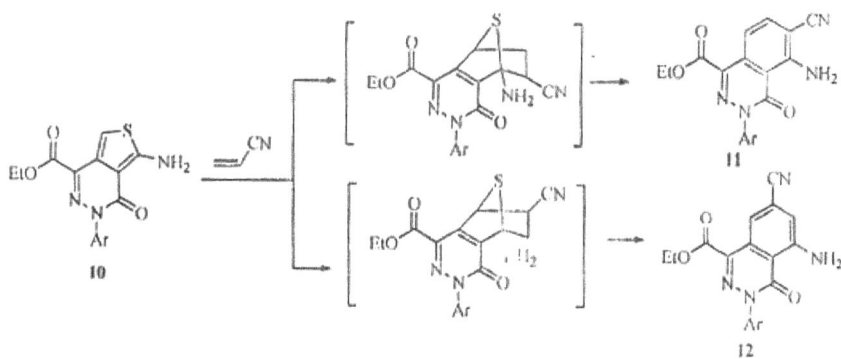

Scheme 3

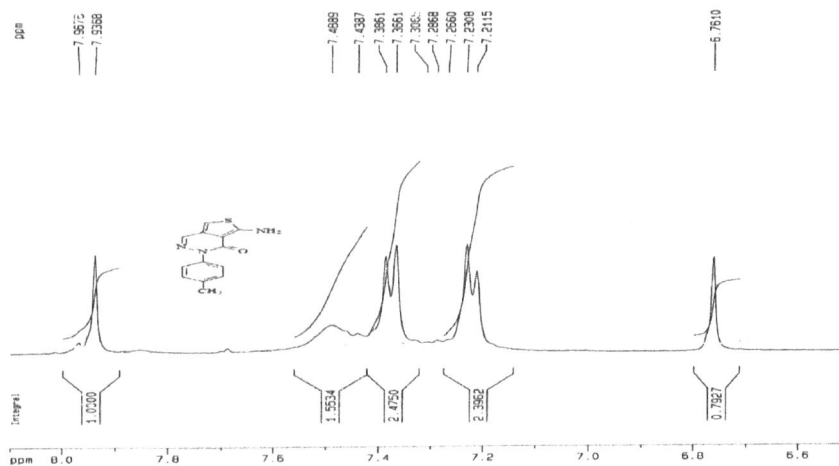

Fig. 4

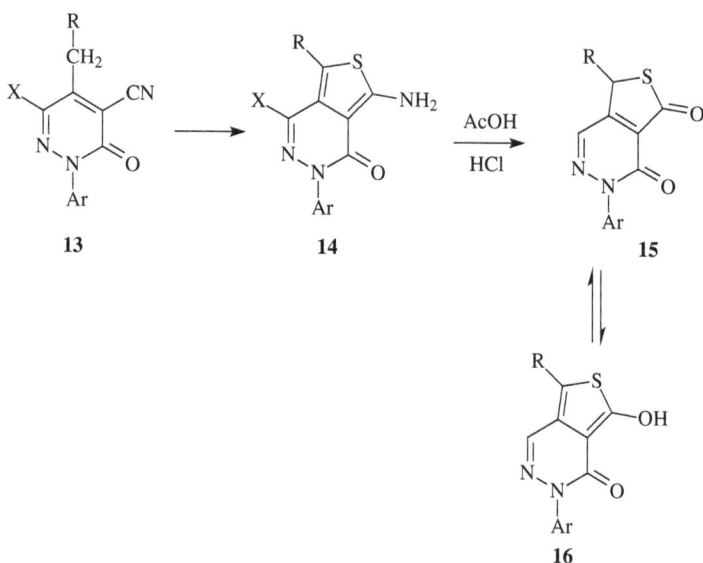

Scheme 4

Hydrolyzing **14** (R = CH$_3$; X = H) affords **15** rather than **16** as indicated by both ^{1}H NMR and ^{13}C NMR (C–H signal at δ = 4.81 ppm and two sp^3 carbons at δ = 43.92 and δ = 20.23). It is possible **16** would not show such signals (Figs. 6 and 7).

Compound **13** (R=X=H) condensed readily with DMF-DMA to yield a product that could be formulated as **17** or **18**; however, ^{1}H NMR revealed two doublets for olefinic protons with J = 12 Hz, attesting that the product formed was indeed **17** in which the protons are in trans. When **17** was refluxed in AcOH/HCl, it afforded pyridopyridazine **19**. Here, ring protons are *cis* and appeared as two doublets at δ = 5.16 and δ = 8.20 ppm with J = 8 Hz (Figs. 8–13).

Compounds **21** and **22** are other examples for enaminones adopting the *E* form and are products formed by condensing **20** and triethylorthoformate (TEO) and piperidine or morpholine. The spectrum (Figs. 14 and 15) showed two doublets at δ = 5.35 and δ = 8.20 with J = 12.8 Hz. The one at δ = 8.20 is for enamine C-1. Pyridazine protons appeared as a singlet at δ = 8.44 ppm (Fig. 14).

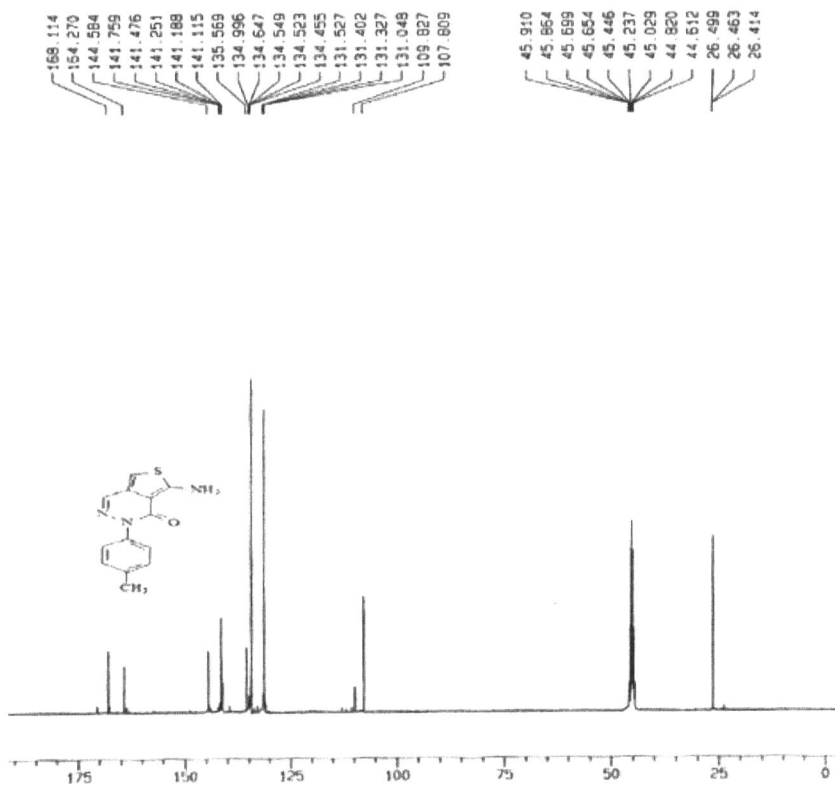

Fig. 5

The ^{13}C NMR revealed that in addition to piperidine carbons, the C-2 in the enamine moiety appeared at $\delta = 89.86$ (Fig. 15).

The same applies for morpholine product **22**. When either **20** or **21** were treated with *p*-chloroaniline, they afforded **23**; in this case, the pyridine protons should reveal a coupling value only at $\delta = 8$ Hz typical for this proton (Fig. 16). In Fig. 17, we show ^{13}C NMR spectra for this product.

Observation of the coupling pattern is important to enable reaching a conclusion regarding the structures in aromatic substitution reactions where electrophiles may attack more than one position in the ring. Thus, reacting an anisole with a cyanoacetic acid–acetic anhydride mixture afforded a cyanoacetylated product, as indicated by MS data and C, H and

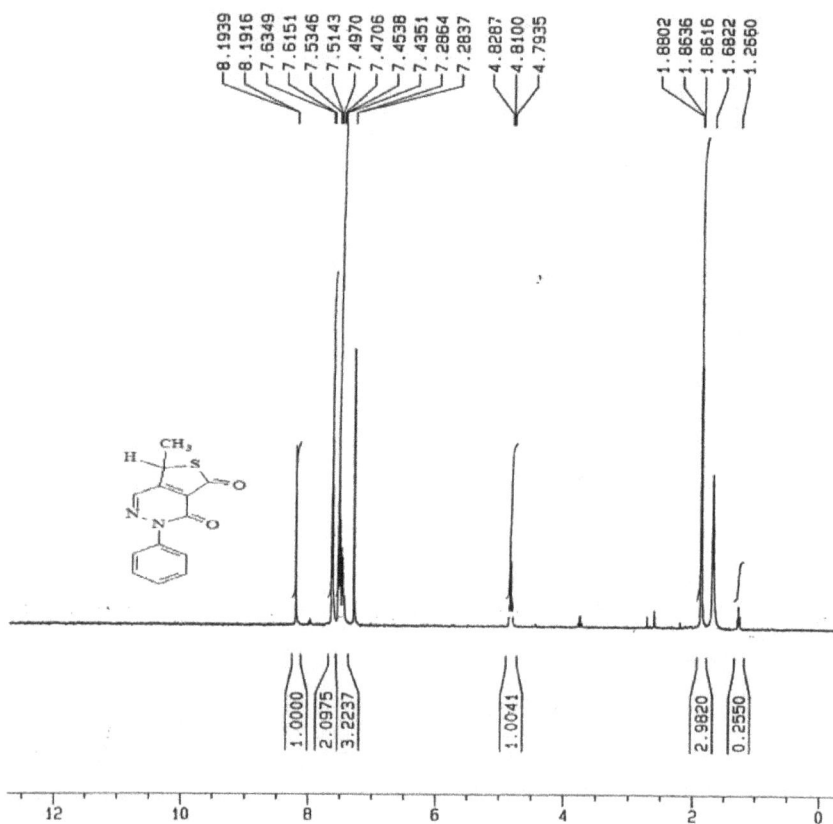

Fig. 6

N analysis. ^{1}H NMR of the product **25** (Fig. 18) indicated that this product is the *p*-substituted compound **25** and not the *o*-substituted one **26**, as was indicated by the two aromatic doublets at δ = 7.2 and δ = 7.6. If the product was **26**, them a completely different pattern should have been observed.

Similarly, cyanoacetylation of **27** afforded **28**, not **29**, as indicated by the coupling pattern. It is noteworthy to report that **28** did not show evidence for the existence of a tautomeric from **30**. Two carbonyl carbons at δ = 176.7 and δ = 173.60 were revealed on ^{13}C NMR in addition to only one *sp*^{3} carbon (CH$_3$) at δ = 26.71. The methylene group (CH$_2$) of the product appeared with a DMSO multiplet at δ = 40.13 (Figs. 19, 20 and Scheme 8).

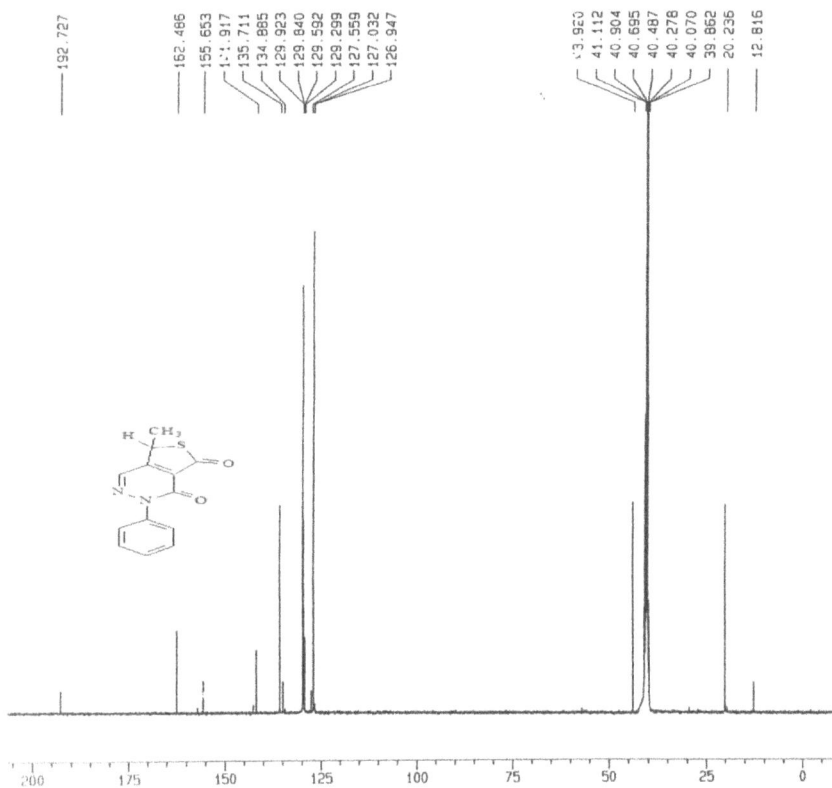

Fig. 7

Simple ^{13}C NMR could resolve a debate about the structure of product formed upon reaction of malononitrile dimer **31** and benzylidenemalononitrile **32**. This product was believed by Elnagdi to be **33**, while Abdelrazek suggested that it was **34**, which had a similar structure. However, ^{13}C NMR (Fig. 21) did not show any signals for carbons linked to sp^3 carbon as would have been the case if the product was **34** (Scheme 9).

The Thiepin-C-l Alkylation Controversy: Coupling constants as conclusive evidence.

In 1994, Prof. Döpp reported the following reaction (Scheme 10).

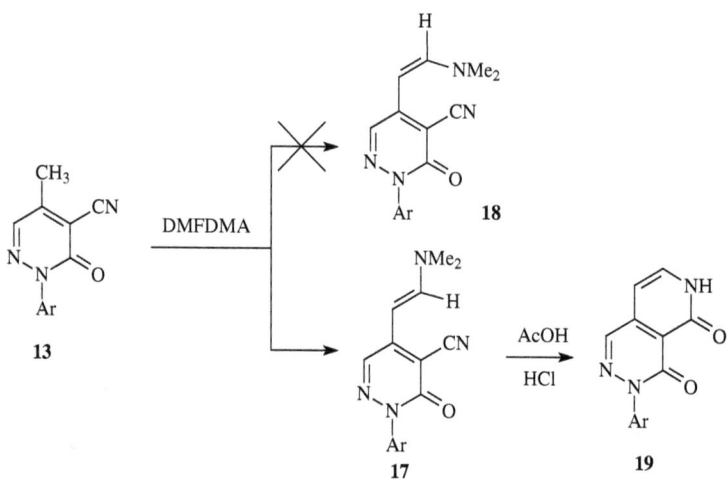

Scheme 5

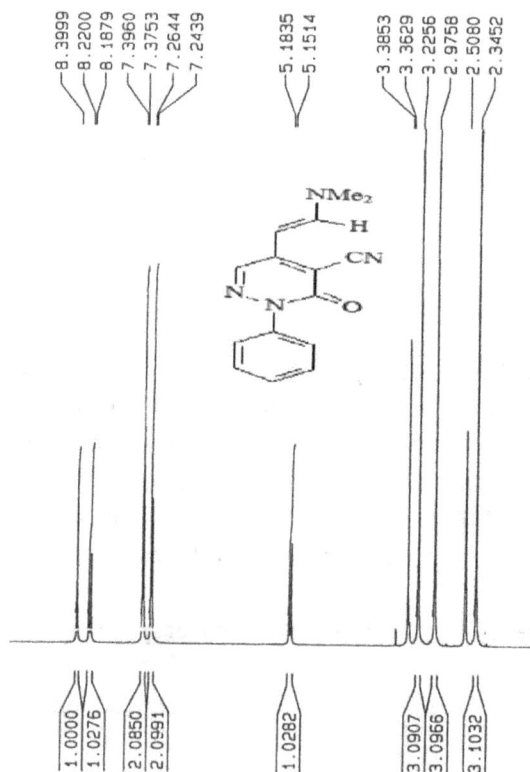

Fig. 8

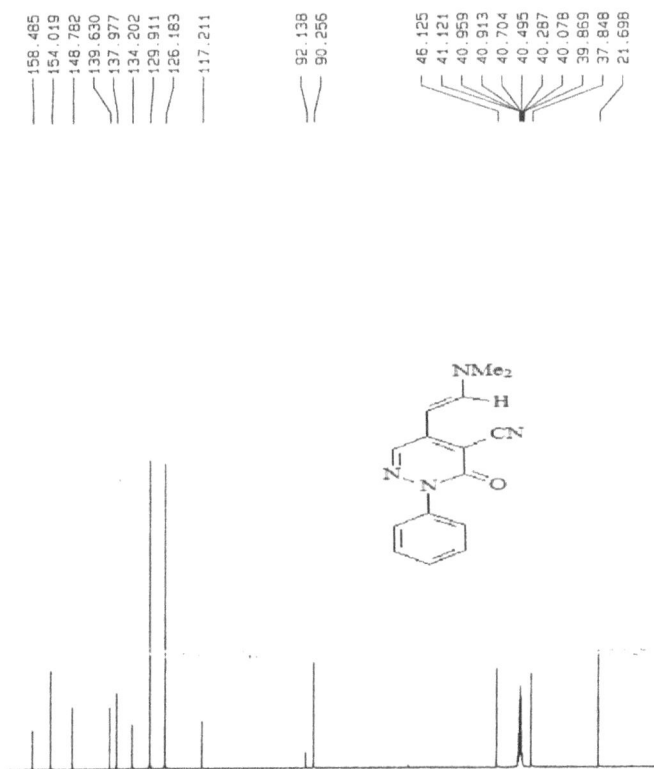

Fig. 9

While the reactions with olefins seemed to parallel our findings, the formation of thermally stable thiepins seemed quite interesting to us. We thus initiated investigation on the corresponding thienocoumarine **16**. While reactions with double bonds proceeded as expected, the reaction with dimethyl acetylenedicarboxylate was found to be temperature dependent. Under reflux, a cycloadduct was formed that loses hydrogen sulfide to afford **18**. This parallels our former results.[7,8] However, when exposed to lower temperatures (cold), a red compound was formed, and we assumed that this also had a thiepin structure[11] (Scheme 11).

Life went on, and during her work on reaction of aminothienocoumarines with enaminones, El-Etaibi and our Kuwaiti M.Sc. student also

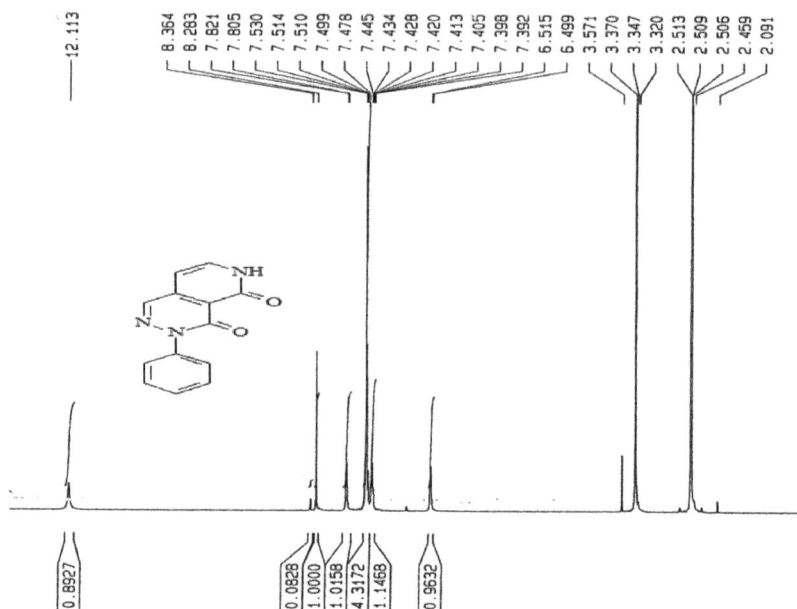

Fig. 10

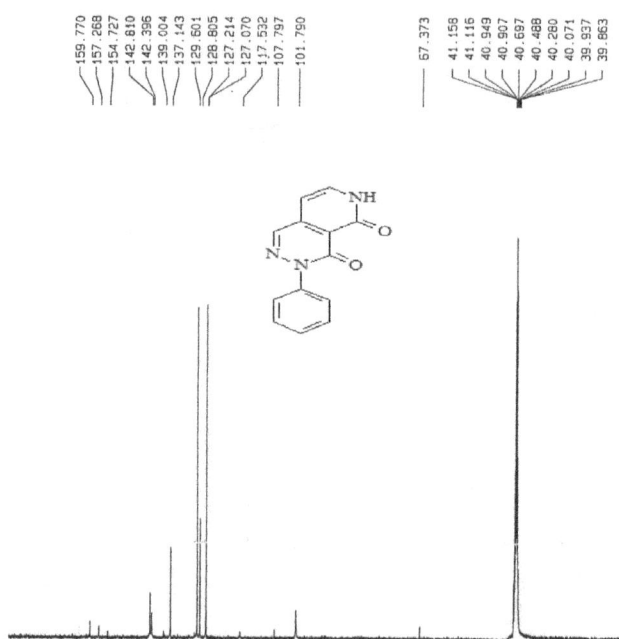

Fig. 11

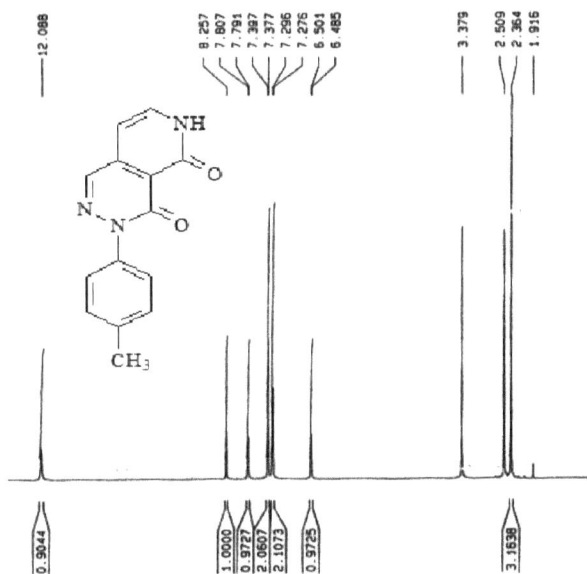

Fig. 12

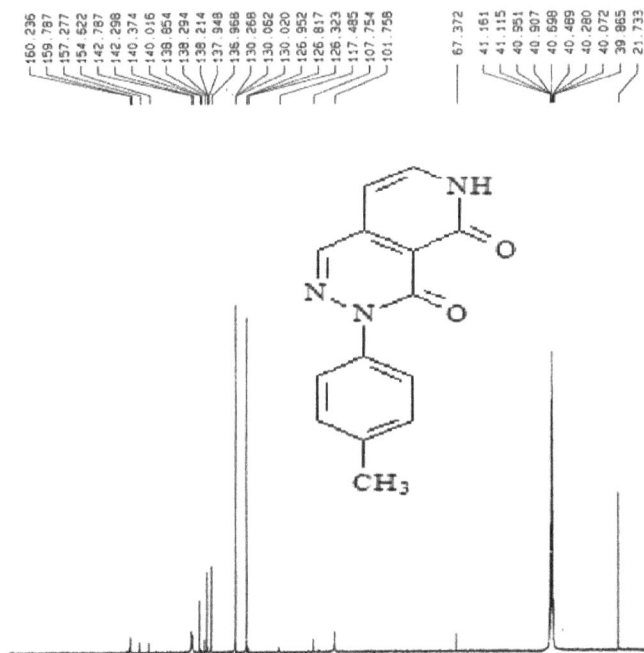

Fig. 13

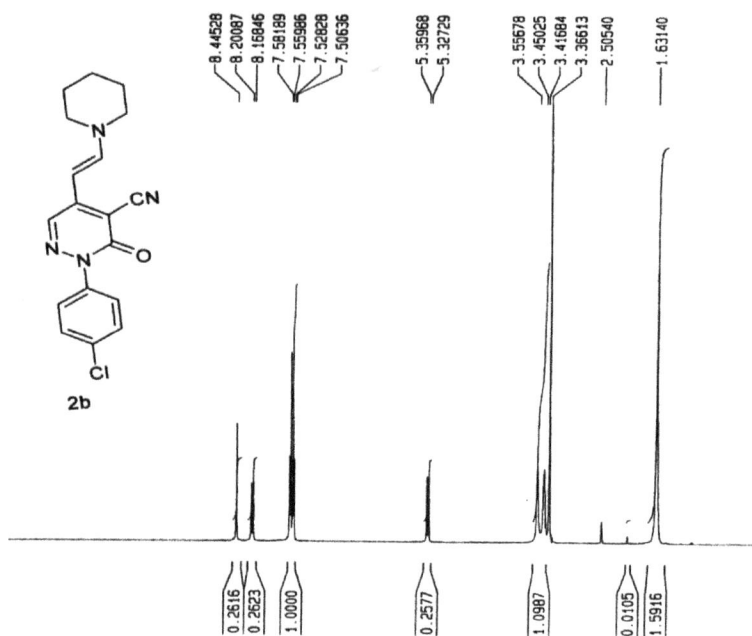

Fig. 14

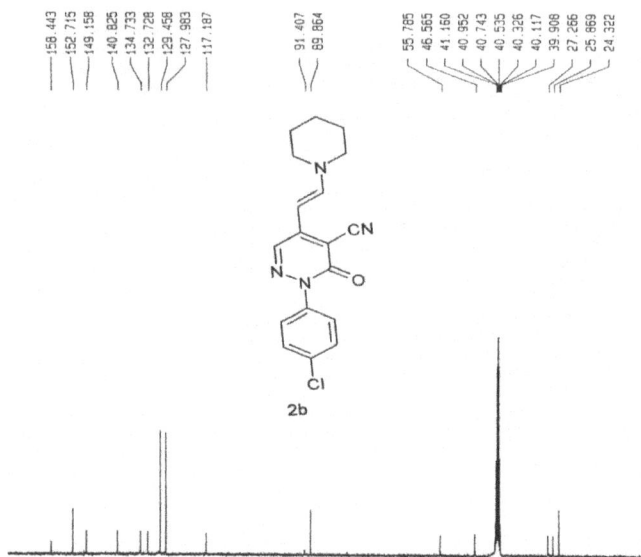

Fig. 15

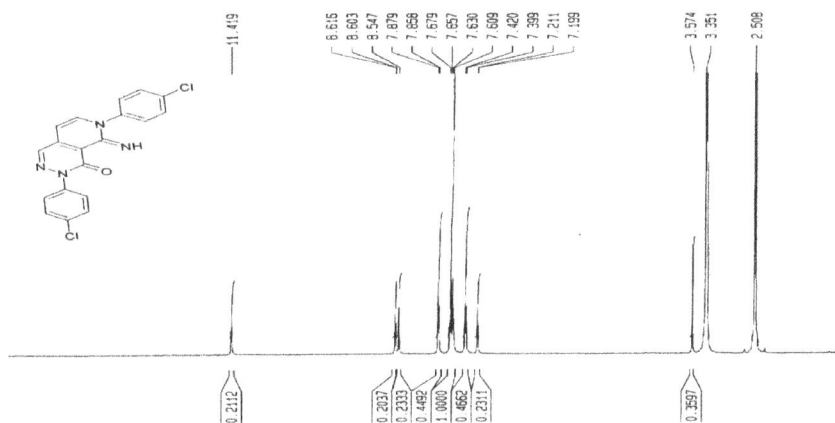

Fig. 16

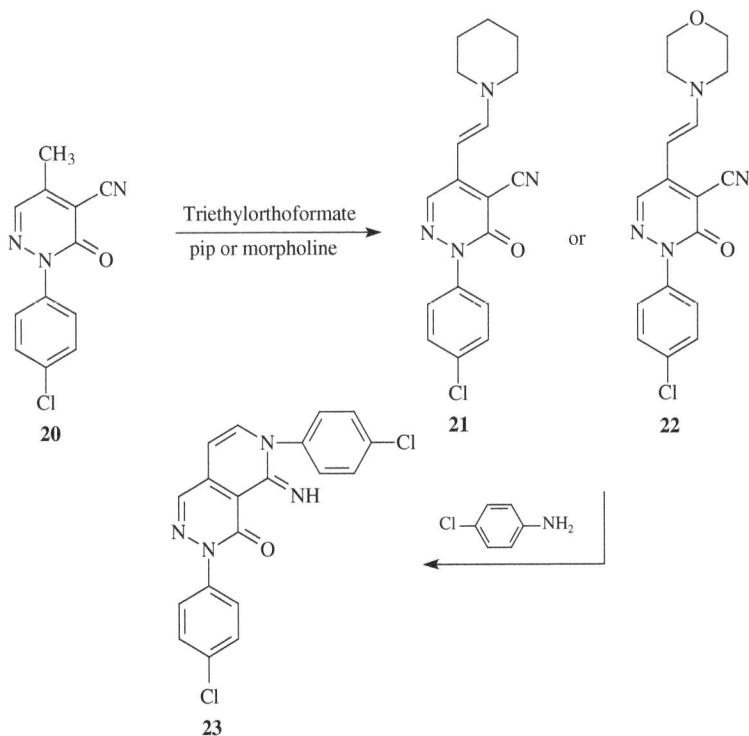

Scheme 6

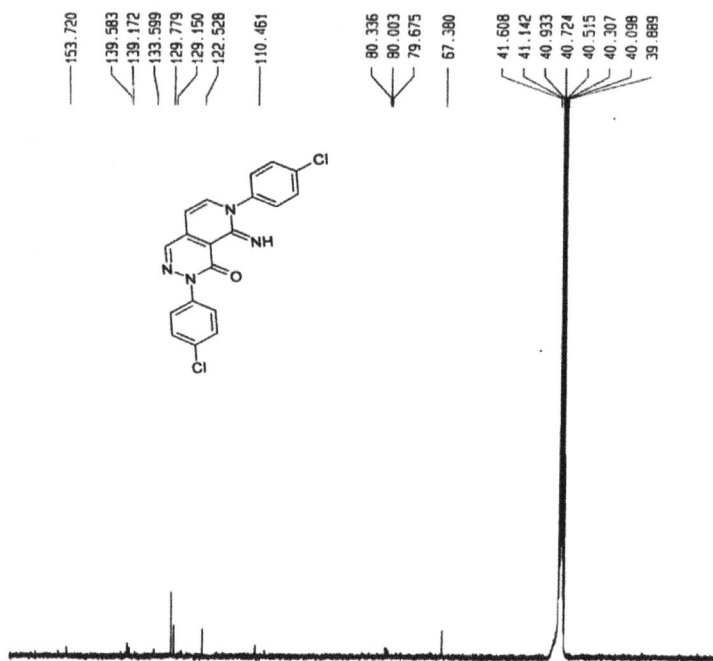

Fig. 17

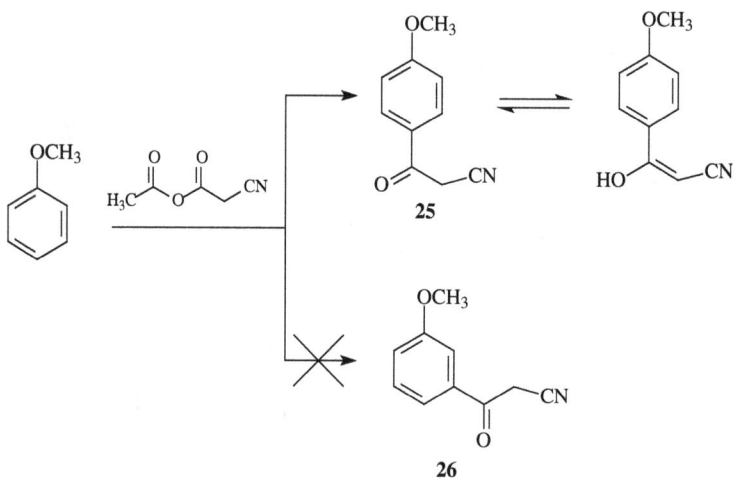

Scheme 7

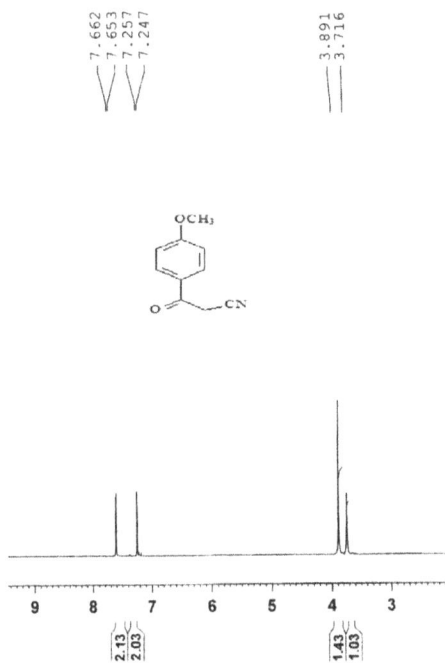

Fig. 18

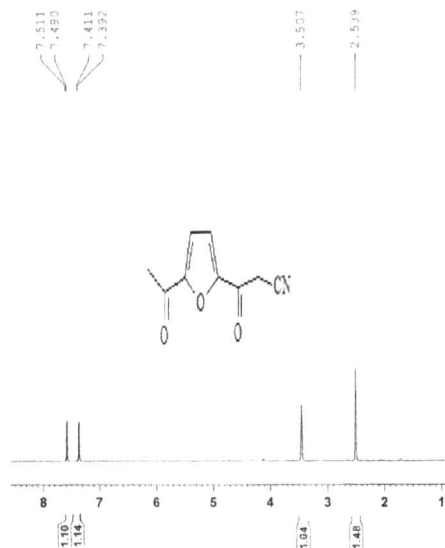

Fig. 19

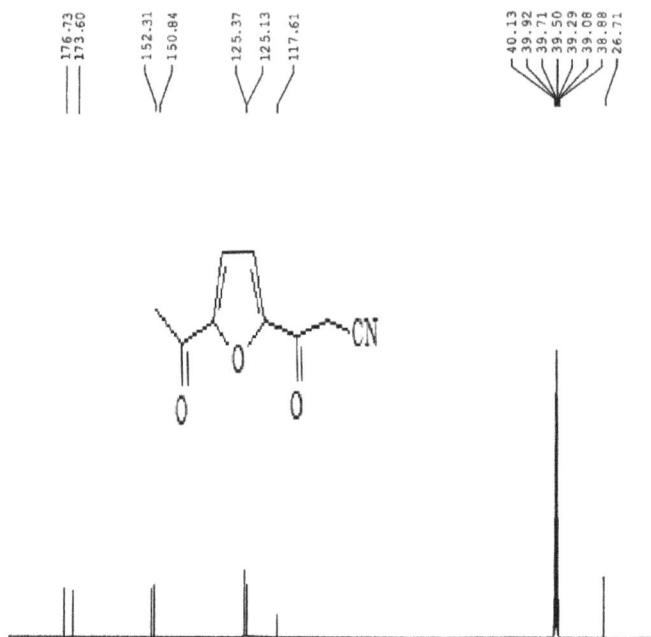

Fig. 20

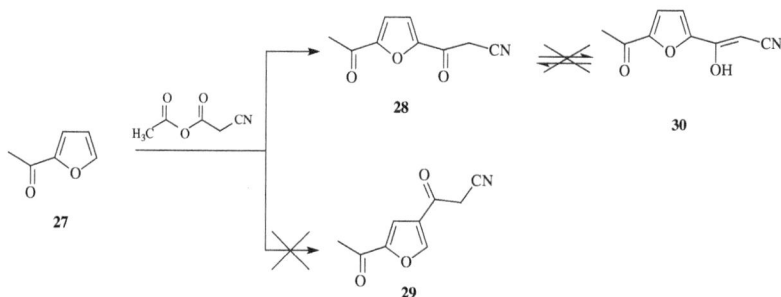

Scheme 8

observed that enaminones condense with aminothienocoumarines under reflux to yield products that did not eliminate hydrogen sulfide. Inspecting the NMR result on the Kuwait University's instrument demonstrated that there was indeed a trans olefinic proton with Ja,b:14 Hz (cf. Fig. 22).[12]

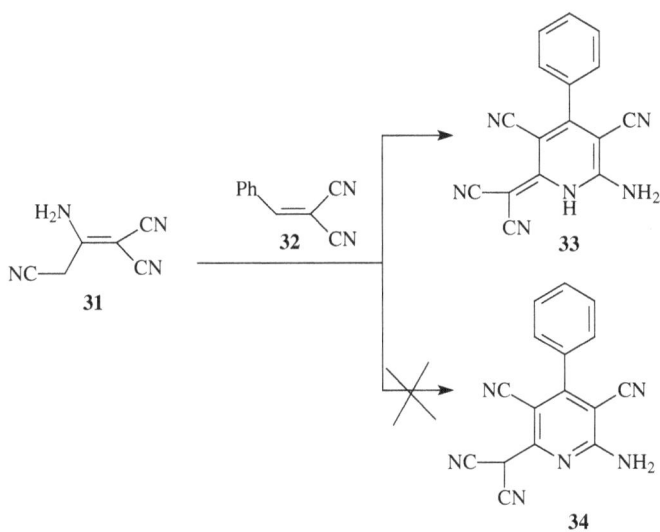

Scheme 9

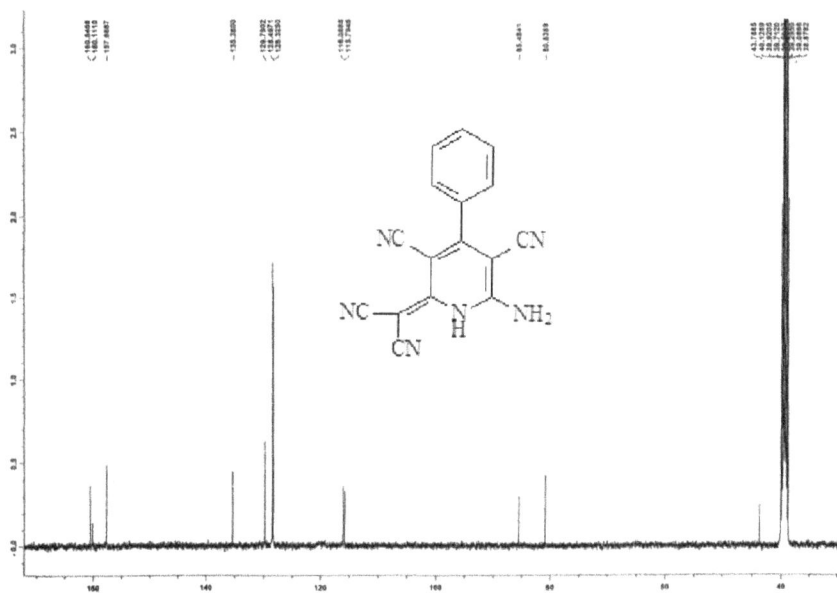

Fig. 21

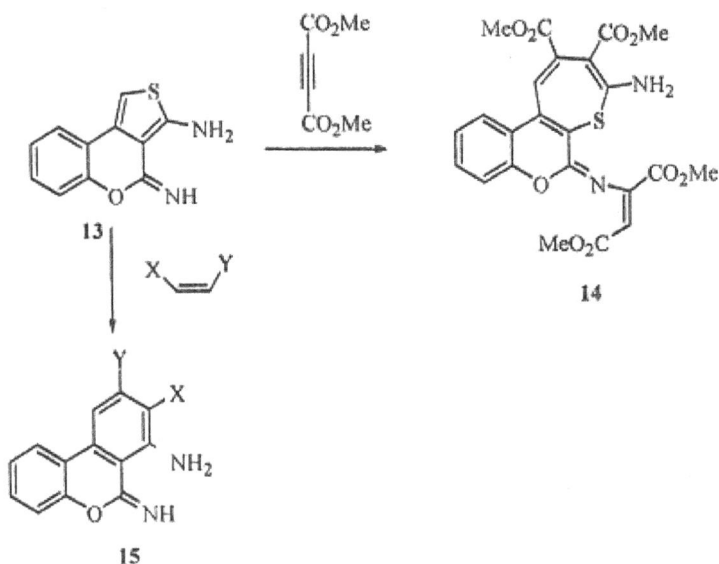

Scheme 10

It then became clear that this could not be a thiepin, as stated in our previous claims, and so a C-1 alkylation structure **20** was assigned to it (cf. Scheme 11). We then reassessed the addition of acetylenes, and this time we added ethyl propiolate and were able to see massive coupling of olefinic trans protons; thus, we revised the thiepin structure and published our corrected conclusions in a subsequent paper.[10] We are glad to see that Prof. Döpp, most likely unaware of the data published by XYZ et al.,[11] also arrived at a similar conclusion that thiepins were never formed (cf. Figs. 22, 23 and Scheme 12).[13]

3.4.2. Thiopyrans or dihydropyridines: Considering magnetic equivalent

Yet another controversy arose when arylidenemalononitriles were reacted with cyanathioacetamide.[14,15] The product that was reported in the literature was dihydropyridine thiols **31** (Scheme 13). However, we obtained ^{1}H NMR and ^{13}C NMR data for the same (with the help of Prof. F. M. Abdelgalil who works at the University of Bonn with W. Steiglish).

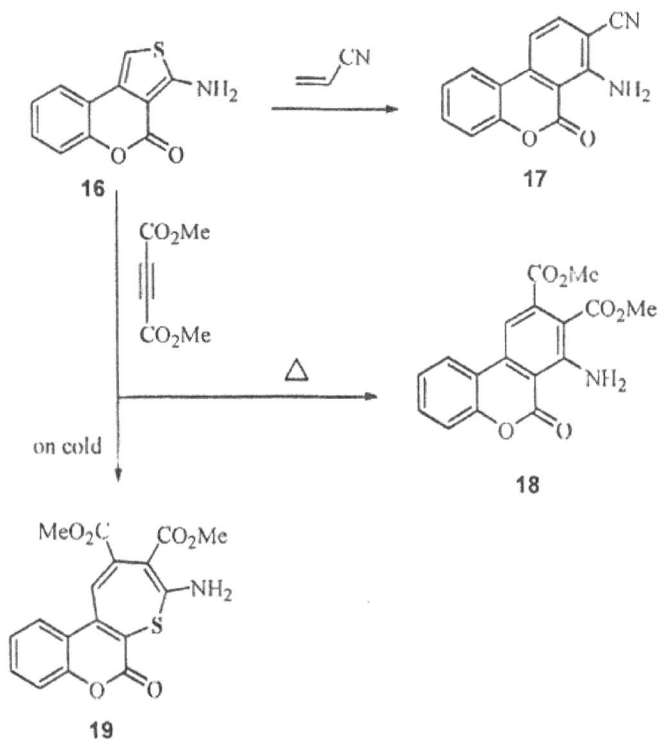

Scheme 11

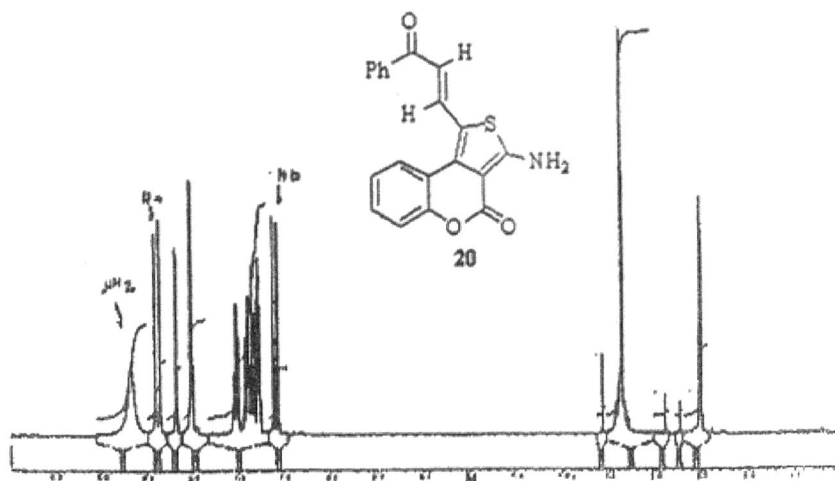

Fig. 22

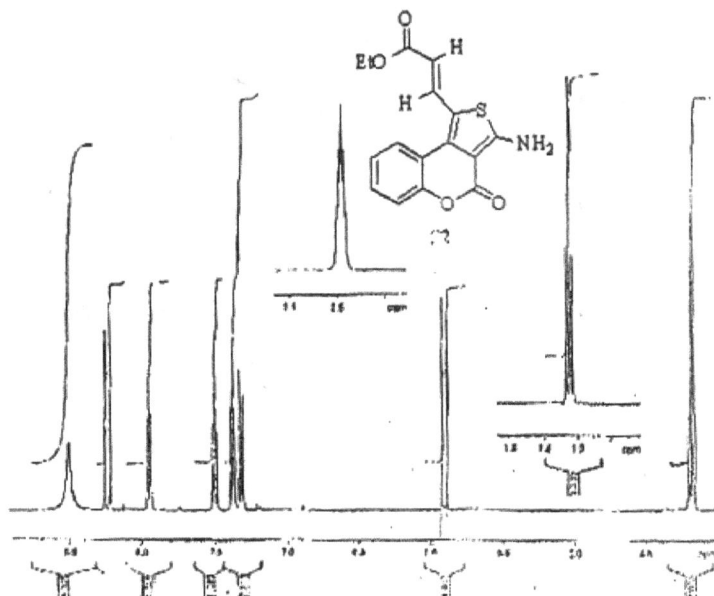

Fig. 23

On first glance, it appeared like a puzzle. Our mind was seemingly was restricted to rationalizing why there was symmetry in this dihydropyridine. It was on a visit to meet Prof. J. Elguero in Madrid (1994), at the Institute de Quimica Medica, when everything became apparent. He took one look at the data and said "You have a thiopyran". The molecule has a plane of symmetry and both the NH_2 groups in ^1H NMR appeared as one signal and in ^{13}C NMR C-2, C-3 and CN carbons were magnetically identical; this is why three carbons were missing from this spectrum (cf. Figs. 24 and 25).

3.5. More Information from One-dimension NMR: DEPT Experiment

Using one-dimension NMR, one can get more information through DEPT and NOE. We will demonstrate the use of both experiments in our work.

First, what is DEPT? It is simply spectral editing to enable defining the number of protons linked to carbons. The result of DEPT experiment

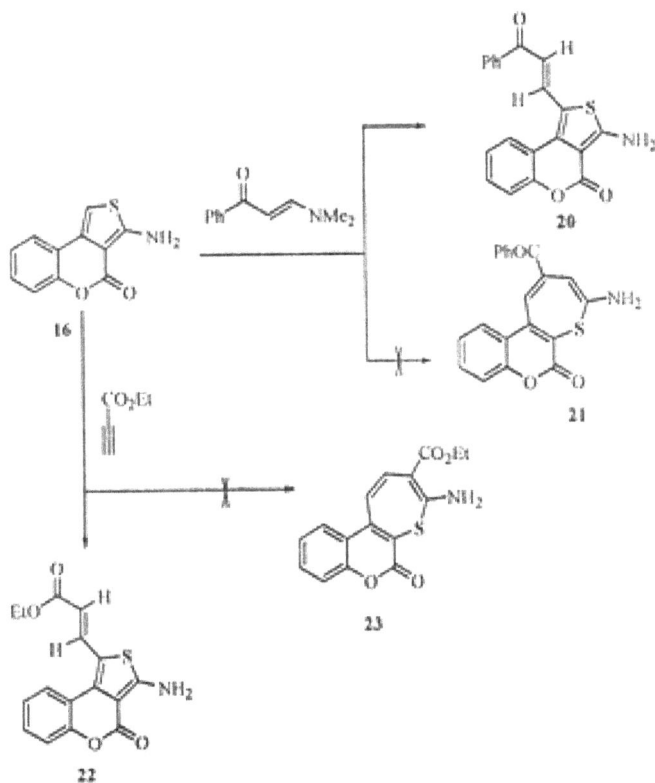

Scheme 12

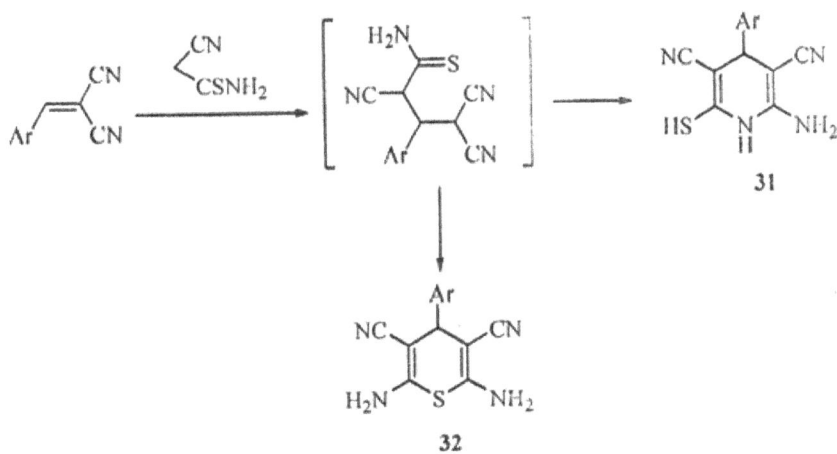

Scheme 13

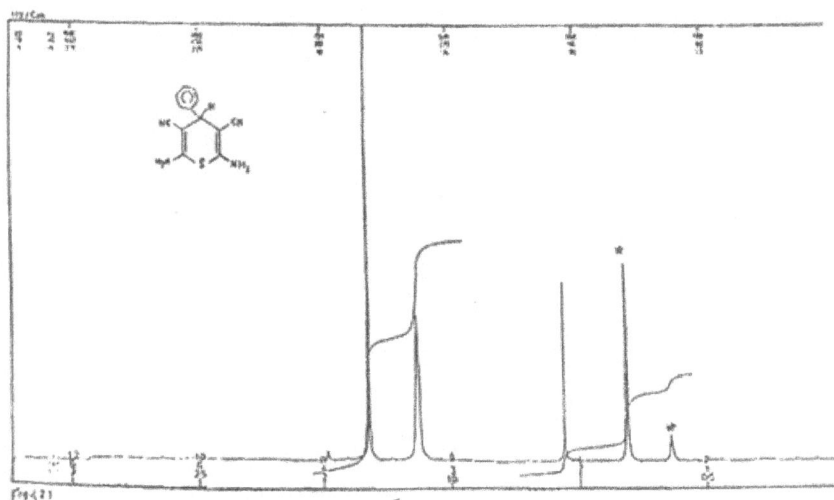

Fig. 24

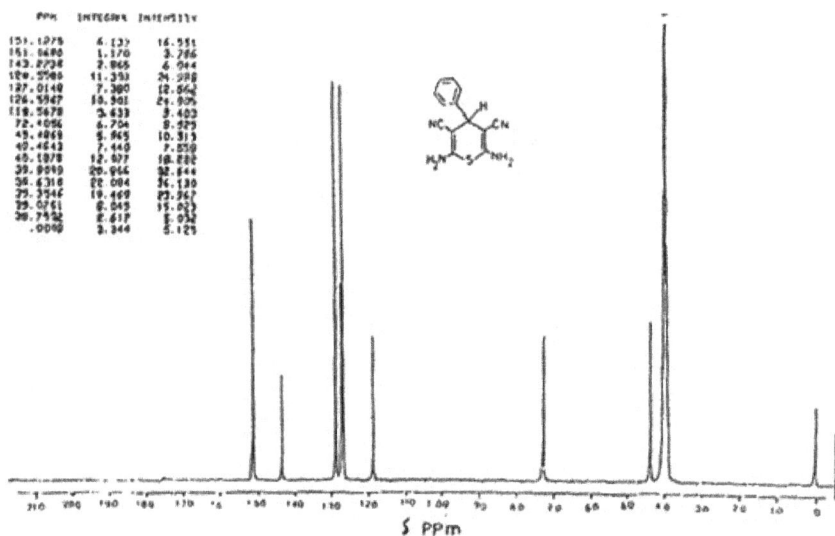

δ ppm

Fig. 25

shows all CH_2s as negative signals and CH and CH_3 as positive signals. This is different from an APT experiment which shows all CH_2s as positive signals and CH and CH_3 as negative ones. During the course of our work, we conducted the reaction shown in Scheme 14.

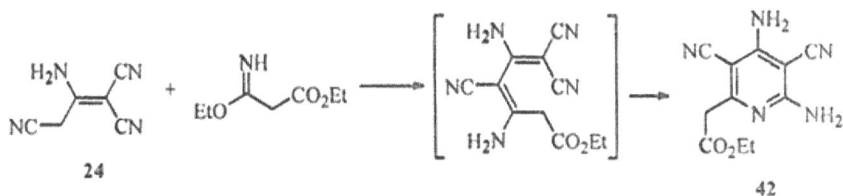

Scheme 14

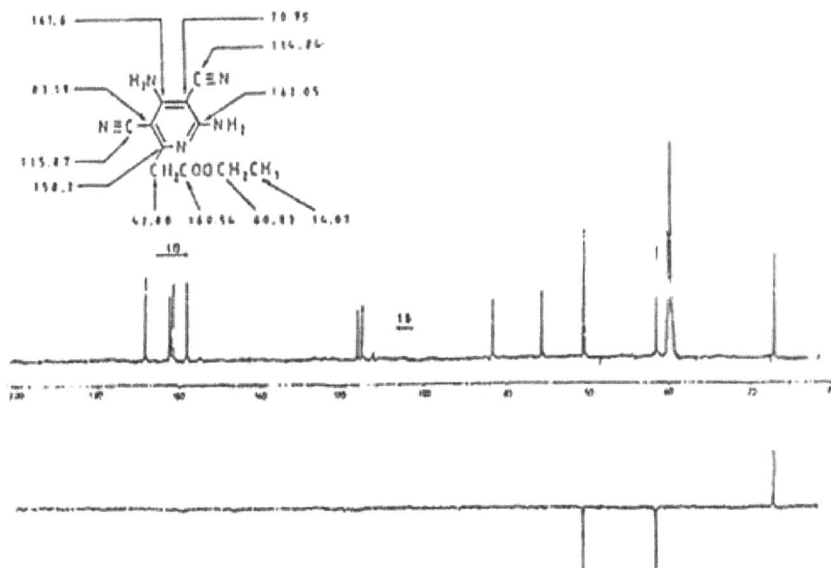

Fig. 26

We obtained a spectrum that was in accordance with proposed structure. Since the product was different from the product that was stated in the literature to have the same structure and was obtained *via* heating malononitrile and ethyl cyanoacetate in presence of a base, we performed a DEPT experiment to further confirm our structural proposal (cf. Fig. 26).[16]

3.6. NOE Difference Experiments

3.6.1. *The origin of nuclear Overhauser effect*

The interaction of one magnetic nucleus with another leading to spin–spin coupling takes place through bonding of the molecules. The information is relayed by electronic interactions, as one can see from dependence of coupling constants on geometrical arrangement of the intervening bond. For example, trans olefinic proton coupling is the largest as efficient overlap of electrons occur (stereo electronic effects).[18] Magnetic nuclei can interact through space, but the interaction does not lead to coupling. The interaction is revealed when one of the nuclei is irradiated at its resonance frequency and the other is detected to have a more intense (positive NOE) or weaker (negative NOE) signal than usual. This is called nuclear Overhauser effect (NOE). The NOE) is only noticeable over short distances, 2–4 Å, and falls rapidly as we approach the inverse sixth power of distance from the nuclei. The effect can normally not be detected by inspection of curves, but the computer subtracts curves before and after irradiation.

3.7. Application to Aminopyrazole Chemistry: Regio-orientation of Addition

In our work, this effect has been successfully employed for decisively solving several research problems. Among these is the regio-orientation of the reaction of 3-metheyl 5-amino-1H-pyrazole (**46**) with ethyl arylidene-cyanoacetate. We noted straightaway that the reaction did not lead, as would be expected from similar studies in the literature, to pyrazolo [1,5-a]pyrimidine as the pyrazole CH at $\delta = 5$–6 ppm was absent. We thus assumed two possible structures (cf. **48** and **49** in Scheme 15).

Full interpretation of data and NOE difference experiments that revealed the enhancement of methyl protons on irradiating aryl ones and *vice versa* are shown in Fig. 27.[17]

3.7.1. Another problem

L-Phenyl-1-oxopropan-3-dimethyl ammonium chloride (**50**) reacted with **46** to afford a product which could either be **51** or isomeric **52**, as ^{1}H NMR indicated involvement of pyrazole H-4 in the reaction. NOE revealed that the product was really **51** as the irradiation methyl signal enhanced the pyridine ring CH-4 at $\delta = 8.3$ ppm and *vice versa* (cf. Scheme 16 and Figs. 28 and 29).[17–20]

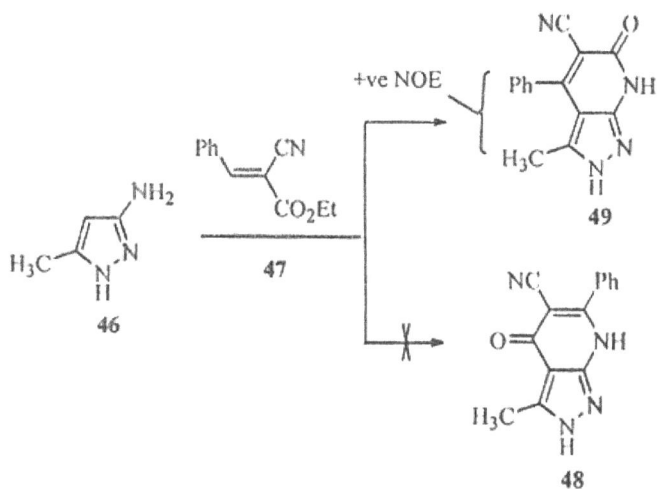

Scheme 15

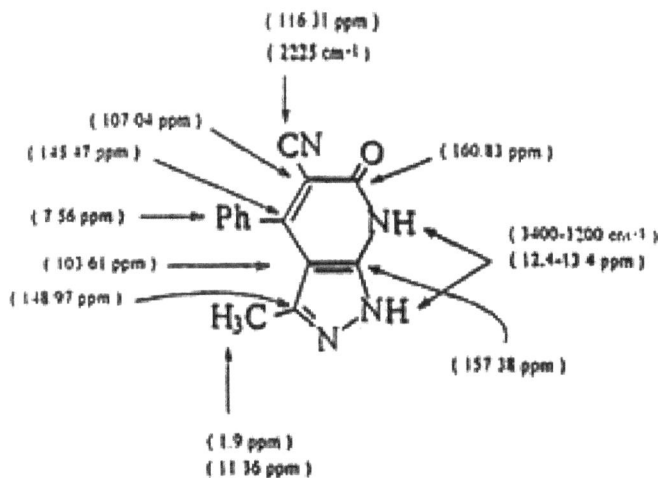

Fig. 27

3.8. Two-dimension NMR Spectroscopy

3.8.1. *Introduction*

The theory of this technique has been detailed in plenty of ways in several textbooks. We will demonstrate only its use. The most important experiments are explained in the following subsections.

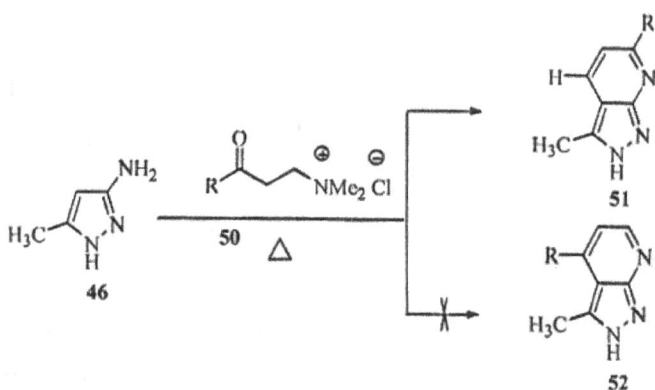

Scheme 16

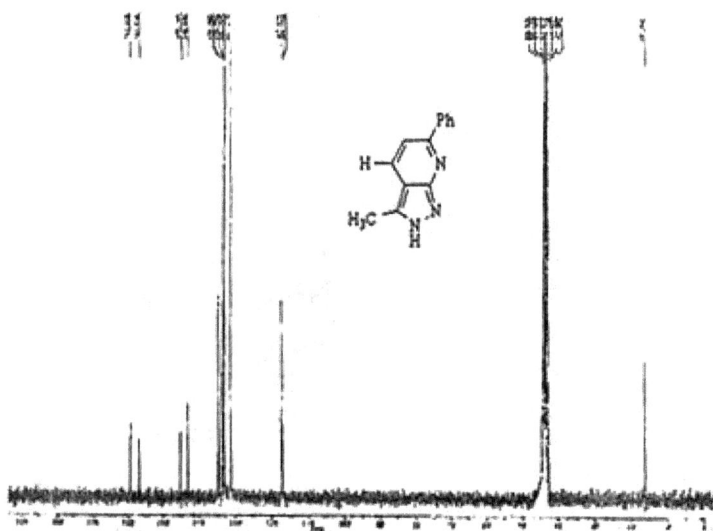

Fig. 28

3.8.2. COSY experiment

In this 2D H,H NMR is drawn, and in this way coupling can be seen easily. It is of course useful for interpreting complex spectra to make things understandable. We will demonstrate the situation by drawing a hypothetical COSY spectra of 2-hexanone. As can be seen, there are open circles and

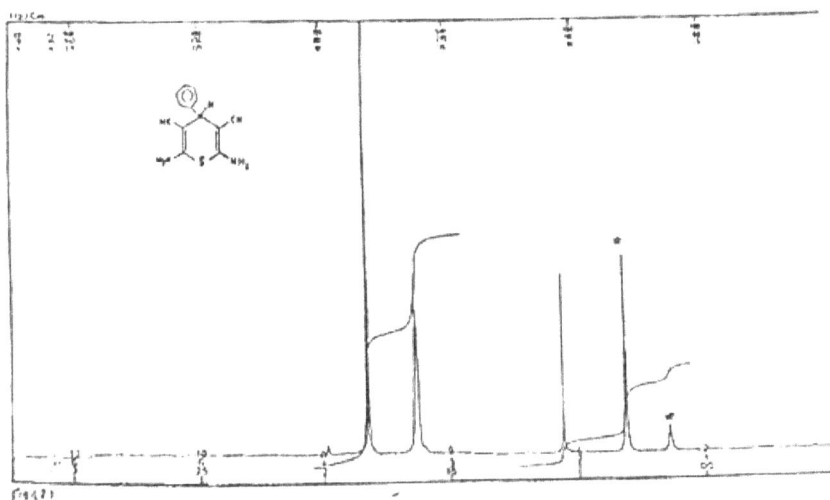

Fig. 29

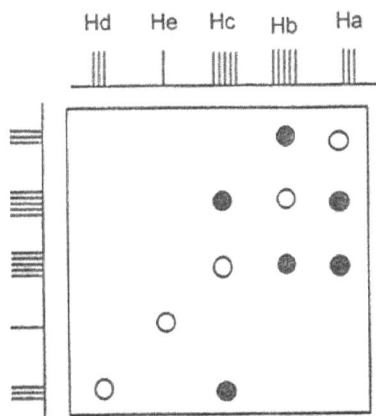

Fig. 30

closed ones. The closed ones are important as they reveal where the open circle (under proton peak) is coupled (Fig. 30).

3.8.3. NOESY experiment

It shows the special arrangement on in other words closed circles appear in form of open ones if protons are close in space (2–4 Å) distance. In the

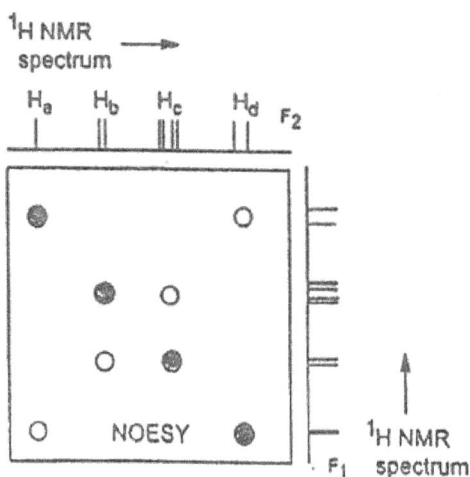

Fig. 31

hypothetical curve, Ha and Hd should be especially proximal while Hc and Hb are also proximal (Fig. 31).

3.9. Application of COSY and NOESY Experiments

It has been recently reported that benzimidazole-2-amine (**53**) reacts with **54a** in acetic acid to yield a mixture of **57** and **60**. In our laboratories, refluxing **53** with **54a** in pyridine solution afforded only one isolable product **57a** at a 70.7% yield. Although the melting point of this product is very similar to that reported for **57a**, the ^{1}H NMR spectrum of the product was found to show some differences. We have identified the reaction product to be **57a** based on elemental analysis, and ^{1}HNMR, COSY and NOESY spectra. Thus, ^{1}H NMR in DMSO-d_6 at +31°C on a Bruker AMX 400 spectrometer showed two triplet-like signals at $\delta = 7.4$ and 7.55 ppm and a doublet at $\delta = 9.6$ ppm, in addition to several overlapping doublets at $\delta = 6.78$, 7.82 and 8.32 ppm. To further confirm the coupling pattern for the six-membered aromatic ring, which is 1,2-disubstituted, we ran a low-resolution COSY 45 spectrum. The two triplet-like resonances at $\delta = 7.55$ and 67.4 ppm are mutually coupled. The respective protons which appeared as multiplets were also identified by COSY spectrum (cf. Figs. 14(a)–14(d)). The $\delta = 7.55$ ppm signal couples to a doublet-like signal at approximately 67.82 ppm, and the signal at $\delta = 7.4$ ppm couples with the one that is approximately at $\delta = 8.32$ ppm. The signals of the phenyl group are almost overlapping these "doublet" signals, and so instead of the using the steady-state NOE experiment, a transient NOE experiment in the form of a 2D-NOESY which was performed in phase-sensitive mode using TPPI and a 300 ms mixing time was used. Positive NOE was observed between the doublet at $\delta = 9.6$ ppm (which is due to H-4) and the multiplet at $\delta = 9.3$ ppm, in which the COSY spectrum should to be coupled to the triplet at $\delta = 7.4$ ppm for the 1,2-disubstituted aromatic ring. Thus, it was concluded that H-4 and H-6 are interacting in space. Therefore, structure **57a** was assigned to this reaction product. Compound **57a** is formed *via* addition of a ring nitrogen to the activated double bond in **54a** yielding a Michael adduct that then cyclizes by losing water and aromatizes *via* losing dimethylamine yielding **57a**. Similarly, the reaction of **53a** with **54b** afforded 57b; we could not isolate a product like **65** which can be formed from intermediates **58** and **59** (Scheme 17 and Figs. 32–35).[21] The comments here were made with the help of Prof. Stotart (a Nobel Prize winner).

3.10. Regio-orientation of Addition of Arylidenemalononitrile

The synthesis and chemistry of pyrazolo[1,5-*a*]pyrimidines have recently been revived as revealed by the vast number of recent papers and patents which report routes for the synthesis of different biologically active substituted pyrazolo[1,5-*a*]pyrimidine derivatives. The recent discovery of Zaleplon (**67**), as an ideal hypnotic drug, has also stimulated further interest in the pyrazolo[1,5-*a*]pyrimidine chemistry. Pyrazolo[1,5-*a*]pyrimidines are readily obtained via reacting bidentate electrophiles with 3(1*H*)-aminopyrazoles. If the reagent is symmetrical or an acyclic intermediate is isolable, then defining the exact structure of the reaction product does not base any significant problem. However, in some cases the only isolable products are the finally formed pyrazolo[l,5-*a*]pyrimidines. In such cases, the identification of the exact regio-orientation of the reactant can be established only with certainty through X-ray crystal structure determinations. In the past, we have described several synthetic approaches to pyrazolo[l,5-*a*]pyrimidine *via* reacting α,β-unsaturated nitriles and esters with 3(*H*)-amino-Pyrazoles,[22] and the structures assigned were mainly

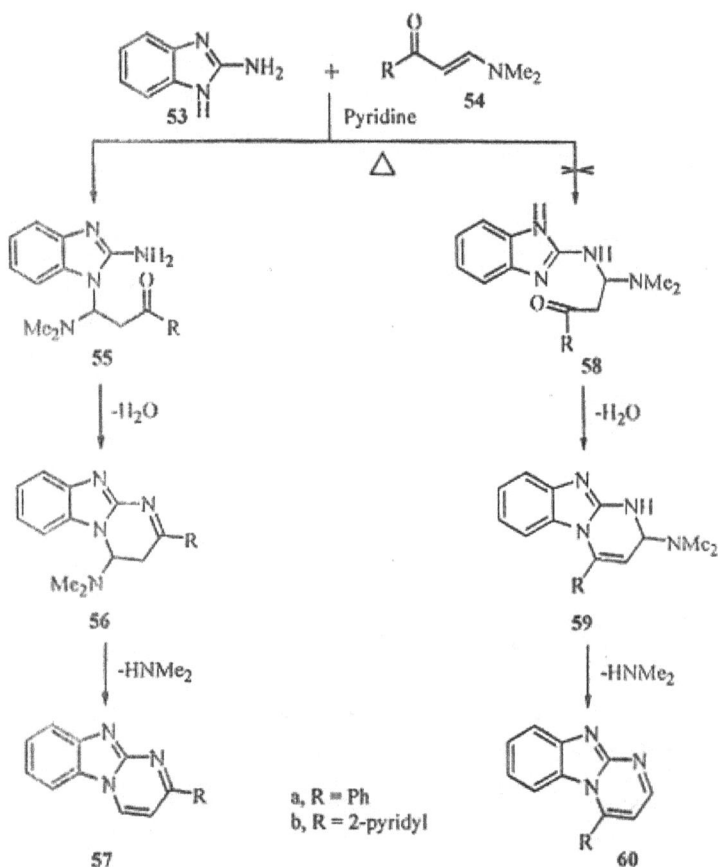

Scheme 17

based on the observed position of amino or carbonyl functions in ^{1}H NMR and IR spectra with the development of 2D NMR techniques. We aimed to develop more conclusive structure elucidation approach via ^{15}N HMBC experiments. A recent report by Meier *et al* has utilized such methodology to establish the site at which electrophiles attack aminoazoles.[23] The aminopyrazole **71a** that has been selected as starting materials could be prepared *via* reacting **68** with hydrazine hydrate in the microwave oven in the presence of acetic acid. On the other hand, compounds **71b–e** were prepared via coupling **69** with aromatic diazonium salts and subsequent

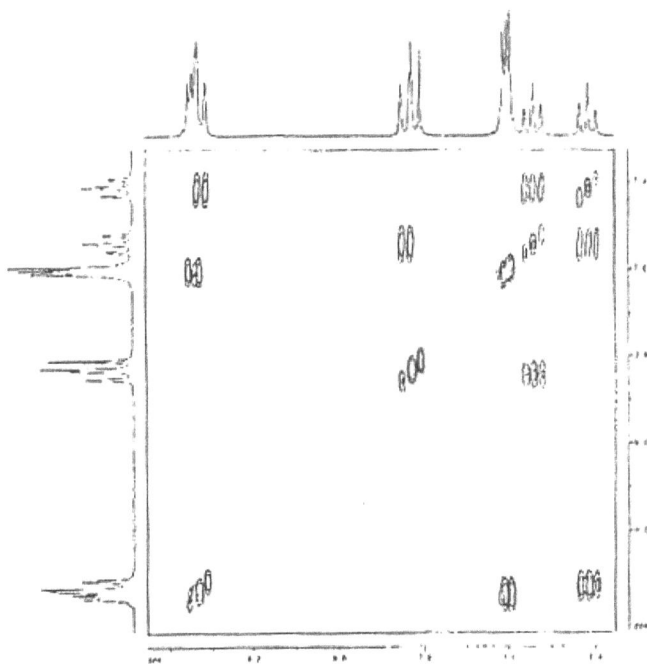

Fig. 32

refluxing of the so-formed arylhydrazones **70a–d** with hydrazine hydrate in ethanolic solutions (Scheme 18).

Compounds **71a–d** reacted with benzylidenemalononitriles **72** to yield aminopyrazolo[1,5-*a*]-pyrimidines that may be formulated as **74a–d** or isomeric **76a–d**. Thus, if the initial addition involved ring nitrogen N-2, as was assumed earlier by Elnagdi *et al*,[28] Michael adduct **73** would be formed. This cyclized and then aromatized to yield **74**. On the other hand, if exocyclic amino function reacts, then **75** would be formed. Its cyclization and auto-oxidation would then afford **76** (Scheme 19). Reacting **71a** with **68** also afforded products that could be formulated as **77** or isomeric **79** (Scheme 19).

Figure 31 shows an ^{15}N, [1]H-heteronuclear multiple bond correlation (HMBC) of compound 77 measured in DMSO-d_6. Cross-peaks for all nitrogen atoms (N-1, N-4, N-7a and 7-NH$_2$) and the protons 2-H, 7-H and NH$_2$ can be observed, provided that they are connected by not more than

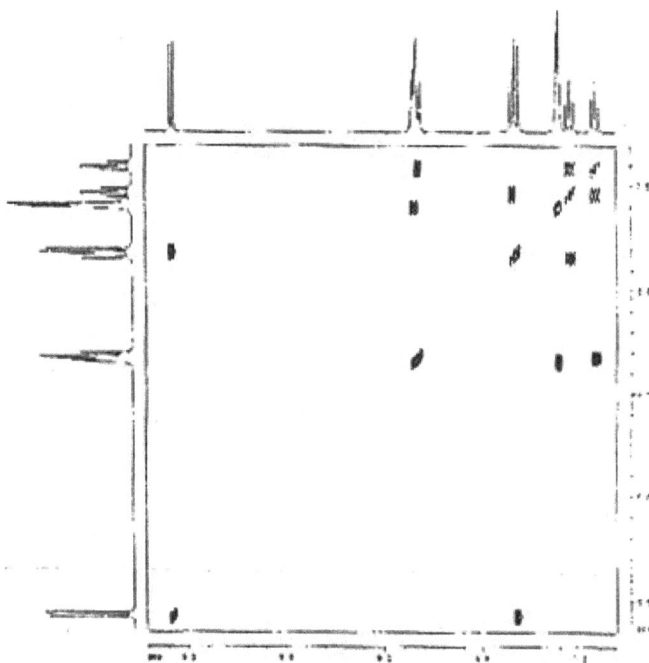

Fig. 33

4 bonds. The size of the coupling constants J (^1H, ^{15}N) corresponds to the sequence $|\,^1J\,| > |\,^2J\,| \approx |\,^3J\,| > |\,^4J\,|$. N-7a with two ^3J and one ^1J coupling gives the largest signal, and N-4 with one ^2J and one ^4J coupling the smallest. The position of the amino group on C-7 is unambiguously determined by the ^3J coupling of its protons with the nodal nitrogen atom N-7a. The coupling ^5J (NH$_2$, N-7a) of the alternative structure (bearing the NH$_2$ group on C-5) would not be visible in the spectrum. On the other hand, large cross-peaks for ^3J (7-H, N-1) and ^2J (77-H, N-7a) should appear in the HMBC spectrum when the isomer with a 5-NH$_2$ group would be present. Figure 1 (^{15}N, ^1H) shows the HMBC spectrum of **71** measured in DMSO-d$_6$. The δ (^1H) values are related to TMS and the A (^{15}N) values to NH$_3$. Additionally, the (^{15}N, ^1H) HMBC spectrum, (^{13}C, ^1H) HSQC and (^{13}C, 1H) HMBC spectra of **77** were measured. The three two-dimensional techniques permit a complete assignment of all ^1H, ^{13}C and ^{15}N signals to certain nuclei of **77** (Fig. 36).

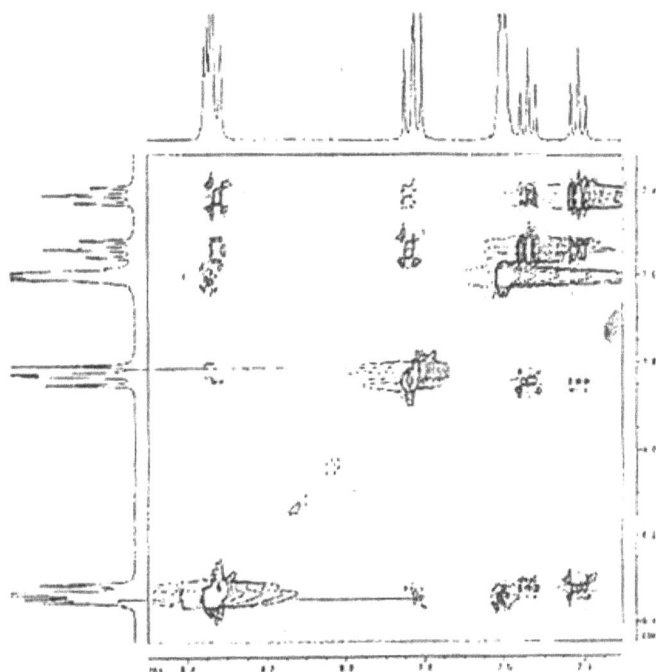

Fig. 34

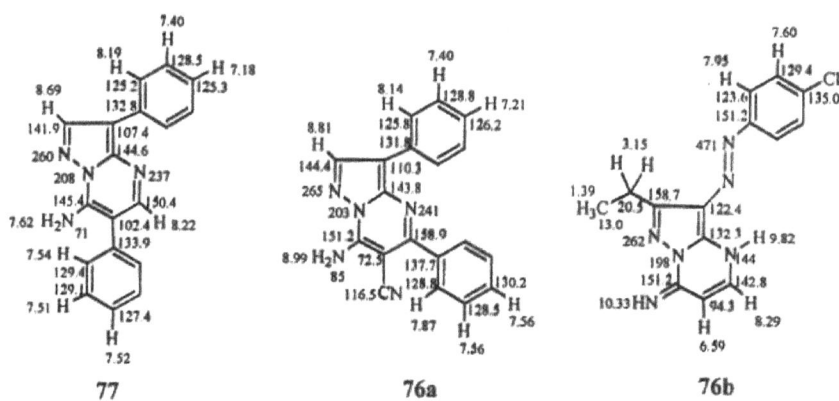

77   76a   76b

Figure 32 shows the-assignment of the ^{1}H, ^{13}C and ^{15}N NMR signals of 3,6-diphenylpyrazolo[1,5-*a*]-pyrimidin-7-ylamine (**77**), 7-Amino-3,5-diphenylpyrazolo[l,5-*a*]pyrimidine-6-carbonitile (**76a**) and 2-ethyl-3-

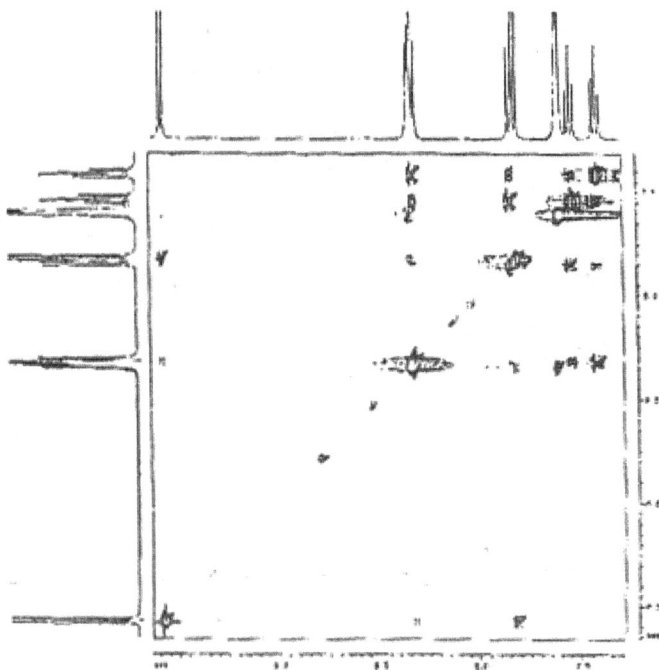

Fig. 35

(4-chrorophenyl)-4H-pyrazolo[1,5-a]pyrimidin-7-ylideneamine   (**76b**). Measurement in DMSO-d$_6$ at room temperature, δ (^{1}H) and δ (^{13}C) values rerated to TMS, (^{15}N) values rerated to NH$_2$. The structure of **76a** corresponds to **77**. Regioselective cyclization yields again a 7-aminopyrazolo[1,5-a]pyrimidine. Figure 2 contains the complete assignment of all ^{1}H, ^{13}C and ^{15}N NMR signals of **76a** based on the three 2D NMR techniques mentioned above. The δ (^{1}H) and δ (^{15}N) values of **76a** and **77** agree very well. A considerable difference exists only in the ^{13}C chemical shift of C-6. The cyano substituent shifts the δ value of this enamine carbon atom to a higher field. All other ^{13}C NMR signals of **76a** are similar to those of **77**. When we applied the three 2D NMR techniques to **76b**, we recognized some new aspects. The ^{1}H NMR measurement in DMSO-d$_6$ revealed molecular dynamics. Compared with **77** and **76**, the NH signals are at lower field in **76b**, namely at δ = 10.33 and 9.82 ppm, and coalesce at δ = 9.94 ppm on moderate warming to 40°C (or enhancement of the

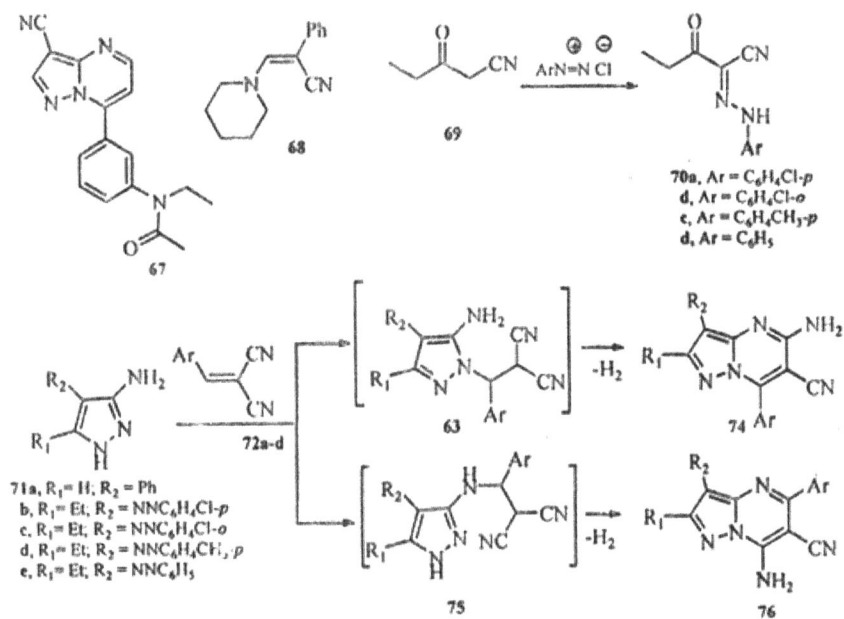

Scheme 18

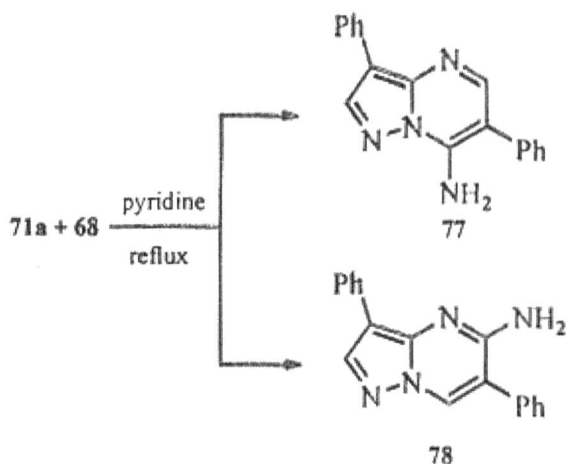

Scheme 19

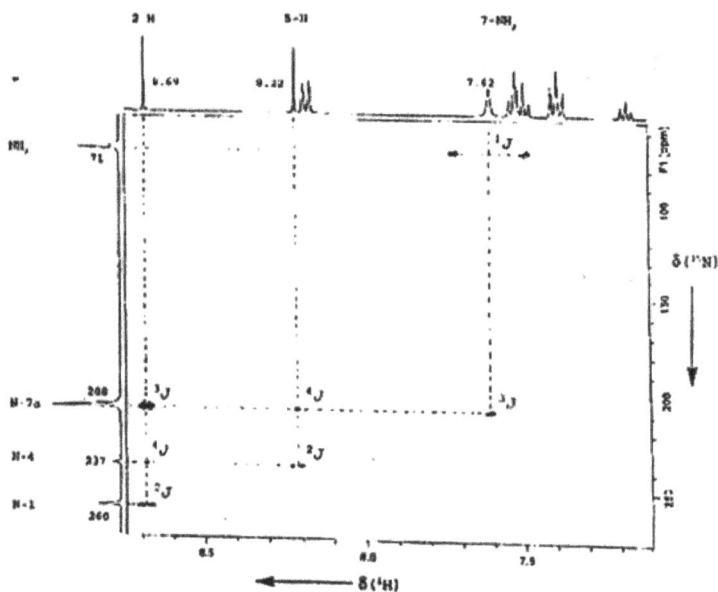

Fig. 36

H_2O concentration). A rotational restriction of the amino group is unlikely. Therefore, we have to consider a tautomeric equilibrium. Scheme 3 shows three possible tautomers **76b**, **76b'** and **76b"**. The (^1H, HMBC, ^{15}N) contains signals for N-1, N-4 and N-7a. Whereas the δ values of N-1 and N-7a are very similar to the corresponding values of **77** and **76a**, N-4 now has a δ value of 144 ppm, which is high-field shifted by almost 100 ppm in comparison to **77** and **76a**. This effect can be explained by a rehybridization of N-4 from sp^2 to sp^3. The (^{15}N, ^1H) HMBC measurement at room temperature in DMSO-d_6 speaks for structure **76b'** as the prevailing tautomer; for **76b**, a δ value of about 240 ppm could be expected for N-4.

The ^{15}N chemical shift of one of the nitrogen atoms in the azo group (Figure 30) can be assigned by the J (^{15}N, ^1H) coupling with the o-H of the benzene ring; its ^{15}N chemical shift of 471 ppm precludes a hydrazono group present in **76b"**. The other nitrogen atom of the azo group cannot be seen in the (^{15}N, ^1H) HMBC spectrum because of a minor polarization transfer. The nitrogen atom of the amino/imino group can also not be seen — due to an exchange mechanism, which includes a Z/E (*syn/anti*) isomerism at this center. Figure **b** shows the assignment of the ^{15}N, ^{1}H and ^{13}C NMR signals of **76b'**.

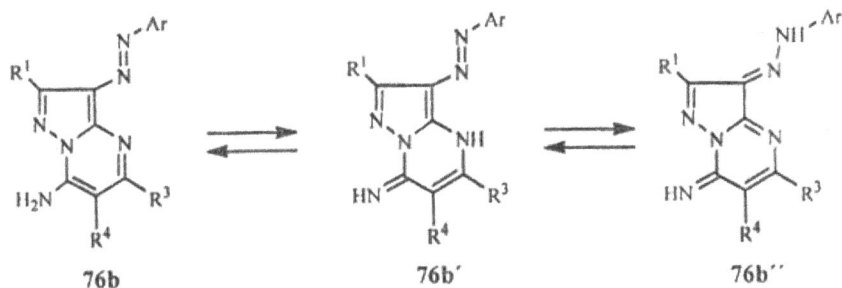

76b   76b′   76b″

We could obtain an X-ray crystal structure **13** upon reacting **72** with **71c**. As clearly indicated (cf. Fig. 37), and contradicting previous beliefs,[10] the reaction product is really 7-aminopyrazolo[1,5-*a*]pyrimidines.

On the other hand, hydrazonals **77** react with dimethyl acetylene dicarboxylate **78** in the presence of triphenylphosphine at room temperature to afford pyradazine **79** rather than isomeric structures **80–82**. The structure of **79** was confirmed by NMR spectroscopy (Scheme 20 and Fig. 38).

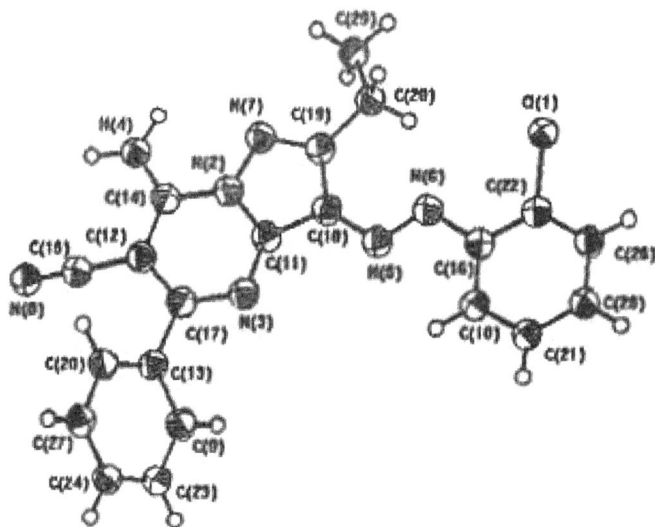

Fig. 37

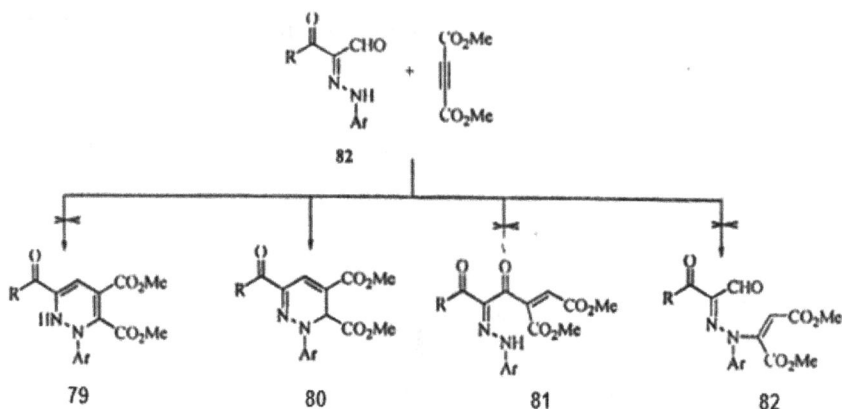

Scheme 20

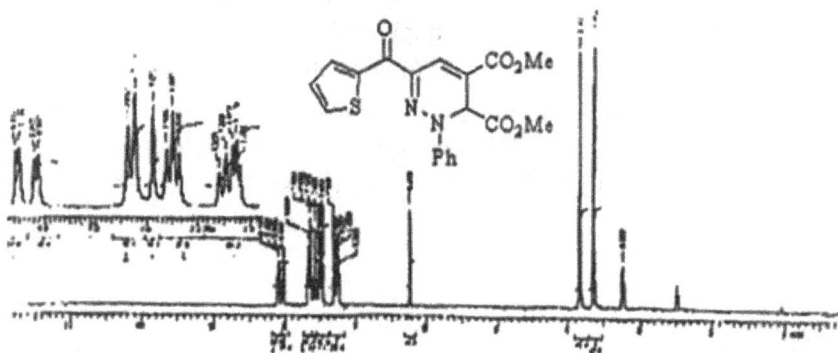

Fig. 38

We encountered another problem while dealing with enaminones as we were trying to add ethyl propionate to **83**. We were expecting 84 or one of its isomers. However, MS indicated that two molecules of ethyl propionate had condensed with one molecule of **83** *via* dimethylamine elimination, and so we considered several isomeric products (**85–87**). Again, simple ¹H NMR enabled us to conclude the correct structure as isomers **86** and **87** as they showed two low-field *o*-protons with large 8–10 coupling. In our spectra (cf. Fig 38), only a singlet was observed, thus confirming structure **85**. A very similar conclusion led to prefer structure

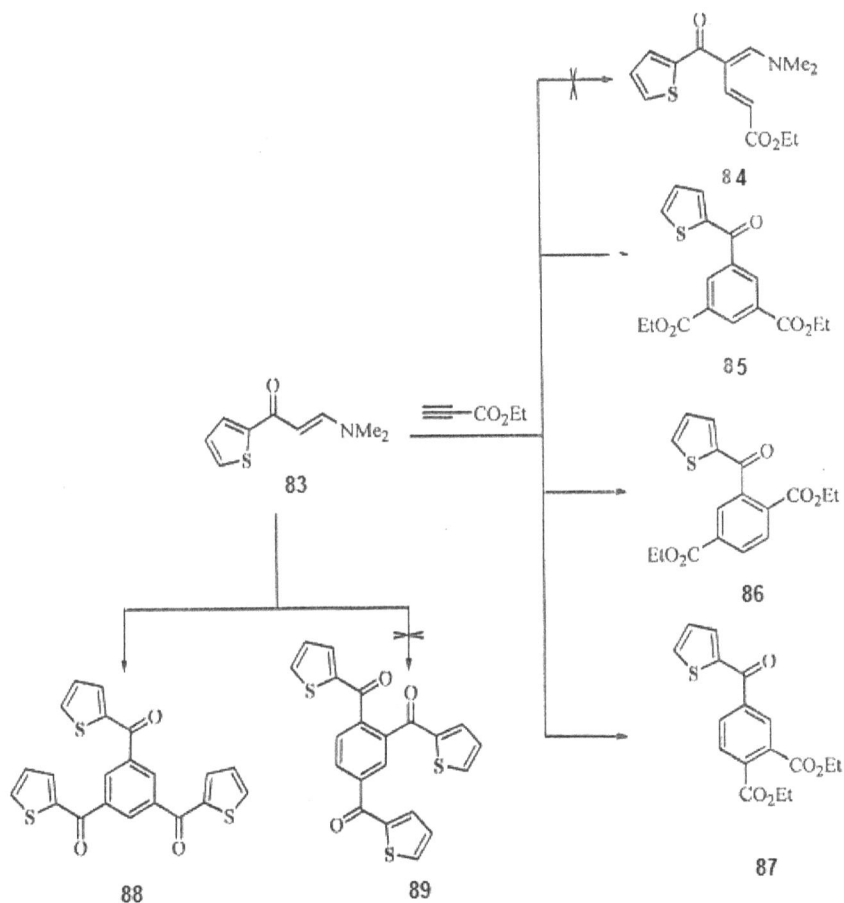

Scheme 21

88 over the possible isomeric **89** for product of condensation of 3 molecules of **87** with elimination of 3 molecules of dimethylamine (Schemes
21, 22 and Fig. 39).

We hope we have offered an easy demonstration of how to use spectroscopy without going in-depth into theories that need knowledge of
physics and mathematics — at least of use to the old generations who has
forgotten that.

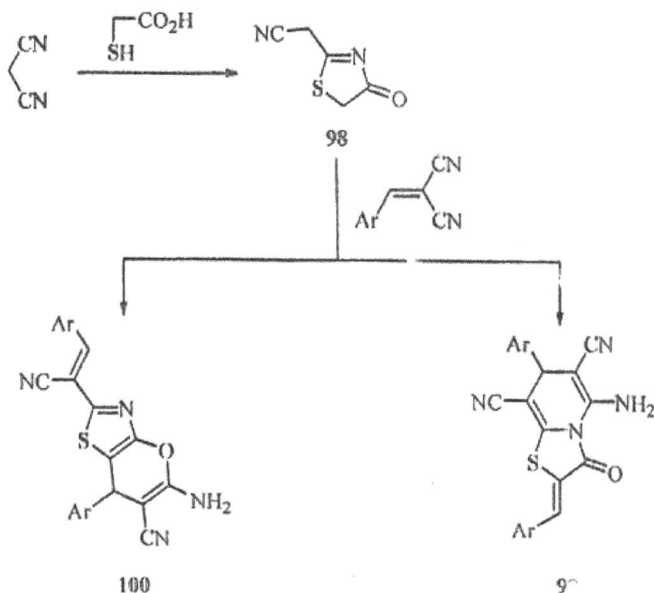

Scheme 22

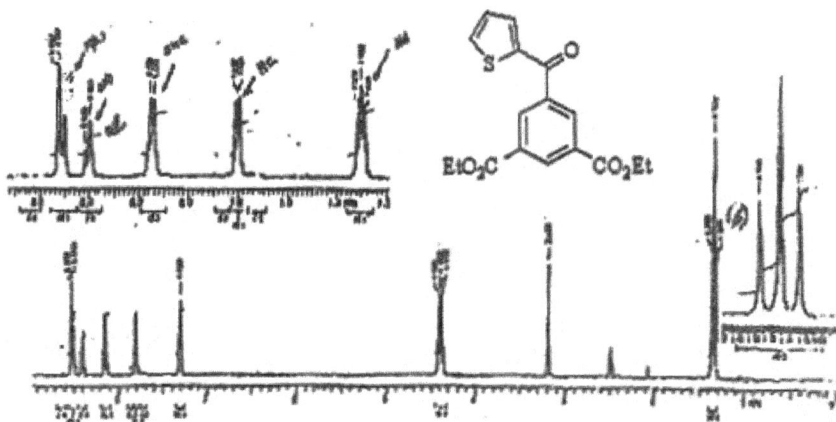

Fig. 39

3.11. Typical Problems Encountered

Problem 1 (solved): In the Ph.D. work of both Dr. I. Abdelhamide, Dr. H. Fakhry, which was recently repeated in the Master's thesis of Mr. Moustafa Sherief Moustafa, the following reactions were performed.

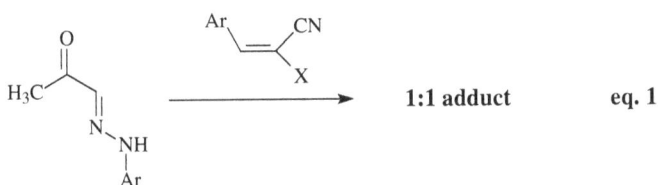

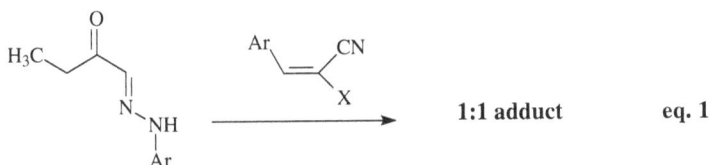

Five structures seemed possible for these adducts (cf. structures **1–5**).

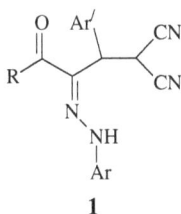

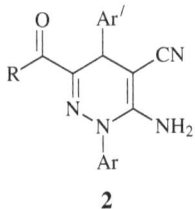

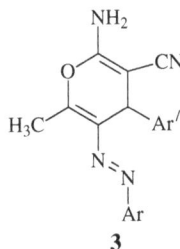

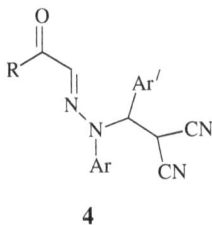

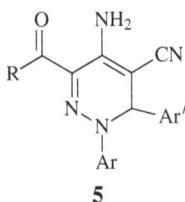

The spectra were obtained for these products of Eq. (1) (Figs. 40–56) by using X-ray to establish structure **2** for the products. However, in fact ^{1}H NMR, ^{13}C NMR and HMBC spectra could also be used, enabling us to arrive to the same conclusion.

Answer: Structures **4** and **1** are easily eliminated based only on ^{1}H NMR that showed amino signal at $\delta = 6.01$ ppm and did not reveal the expected doublets for protons linked to sp^{3} carbon. ^{13}C NMR clearly showed the carbonyl carbon at $\delta = 195$ ppm; thus, the pyran structure could also be eliminated (structure **3**).

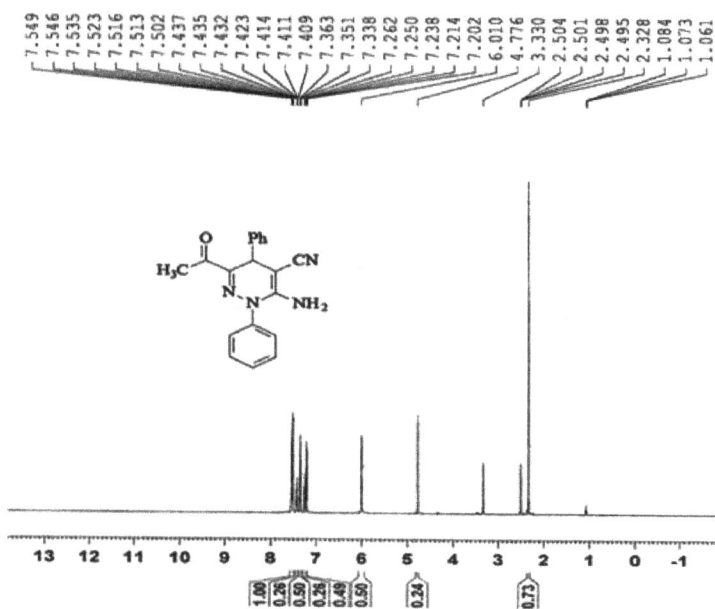

Fig. 40

In HMBC experiment, the carbonyl carbon in the product showed a cross-peak with one proton signal at $\delta = 4.77$, confirming the structure as **2** — the spectra obtained for the product of Eq. (2) is shown in Figs. 44–50. The main features observable in the spectra appeared. However, the CH_2-CH_3 appeared as two separate quartets. The student in fact believed initially that there was something.

Wrong, but the NMR specialist in the department informed him that the protons of CH_2 are prodiastomeric as the system also contains another chiral center and thus both protons are magnetically different, thus leading to the observation of two separate quartets.

Problem 2: Figures 57–70 show data obtained for the product of the reaction in Eq. (3), and Figs. 26–31 show the data obtained for the product of the reaction in Eq. (4). The same analytical methods used for deducing the structure of products shown in Eqs. (1) and (2) were used to deduce the structure of those products.

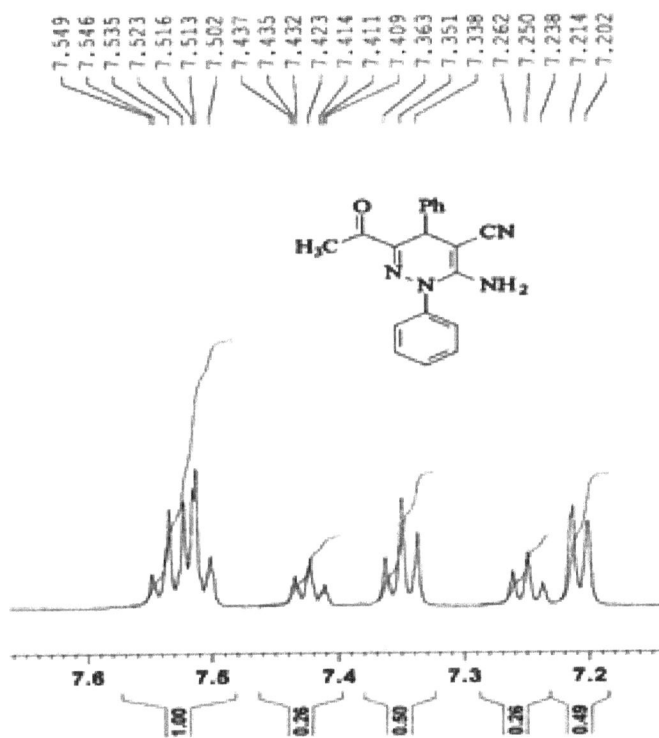

7.549
7.546
7.535
7.523
7.516
7.513
7.502
7.437
7.435
7.432
7.423
7.414
7.411
7.409
7.363
7.351
7.338
7.262
7.250
7.238
7.214
7.202

Fig. 41

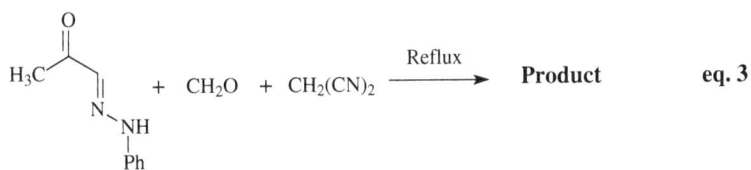

$$H_3C \quad + \quad CH_2O \quad + \quad CH_2(CN)_2 \xrightarrow{\text{Reflux}} \textbf{Product} \qquad \textbf{eq. 3}$$

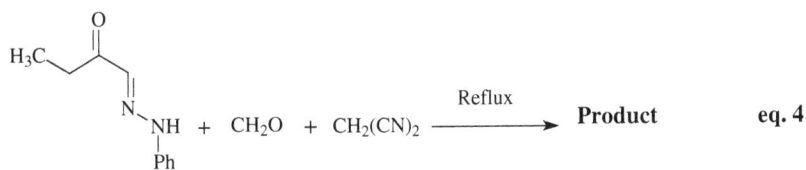

$$H_3C \quad + \quad CH_2O \quad + \quad CH_2(CN)_2 \xrightarrow{\text{Reflux}} \textbf{Product} \qquad \textbf{eq. 4}$$

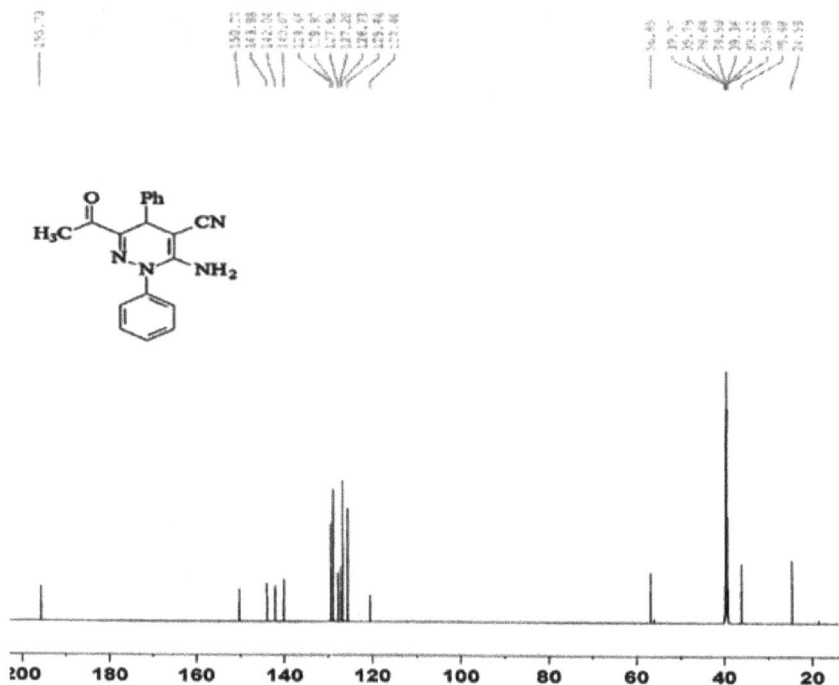

Fig. 42

Problem 3: The reactions in Eqs. (5) and (6) definitely yield the given products. Use the data from Figs. 71–77 to interpret these structures. Discuss why carbonyl appears in the indicated positions at $\delta = 160$ ppm and at $\delta > 180$ ppm as typical C=O.

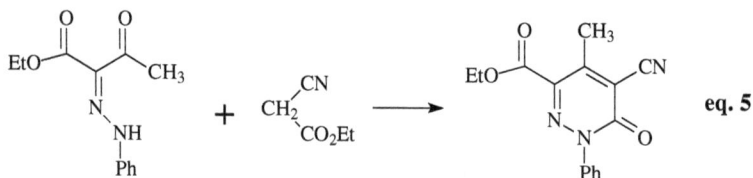

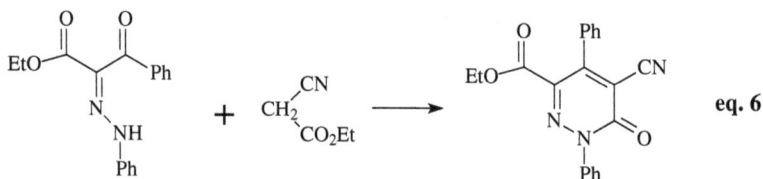

MS-AZ29 #128 RT: 5.63 AV: 1 NL: 9.22E7
T: + c EI Full ms [49.50-900.50]

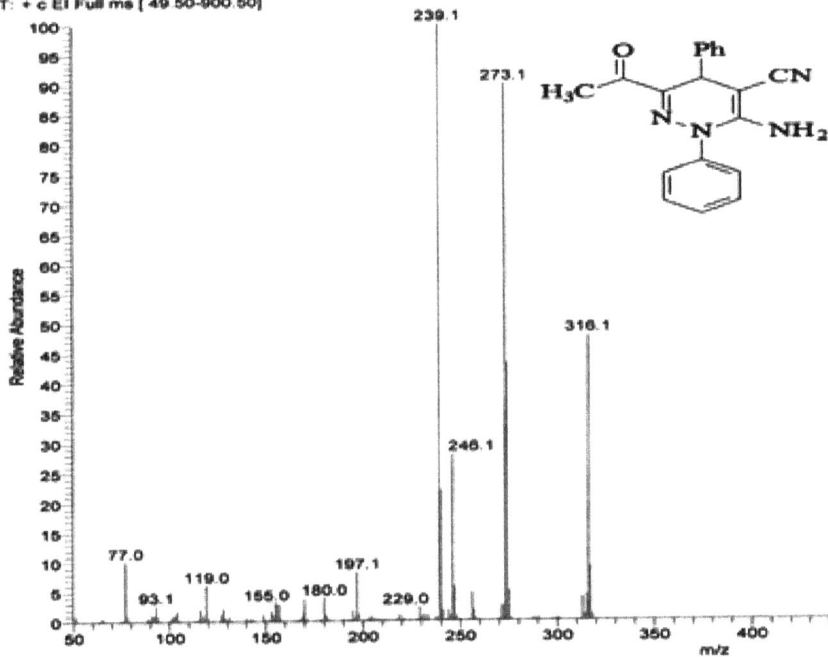

Fig. 43

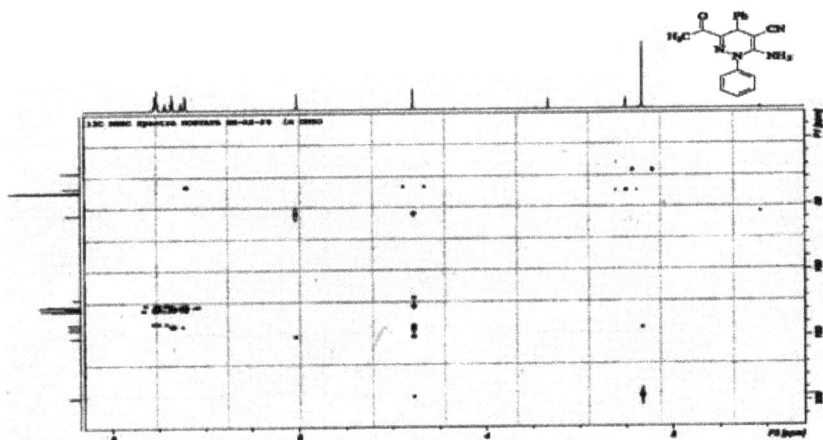

Fig. 44

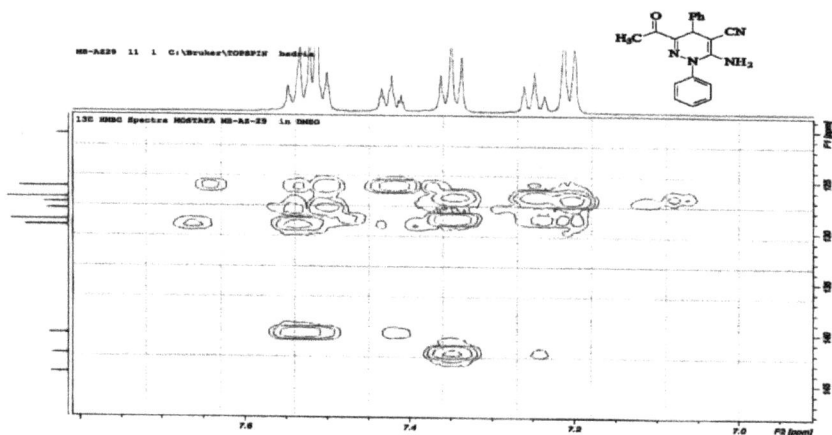

Fig. 45

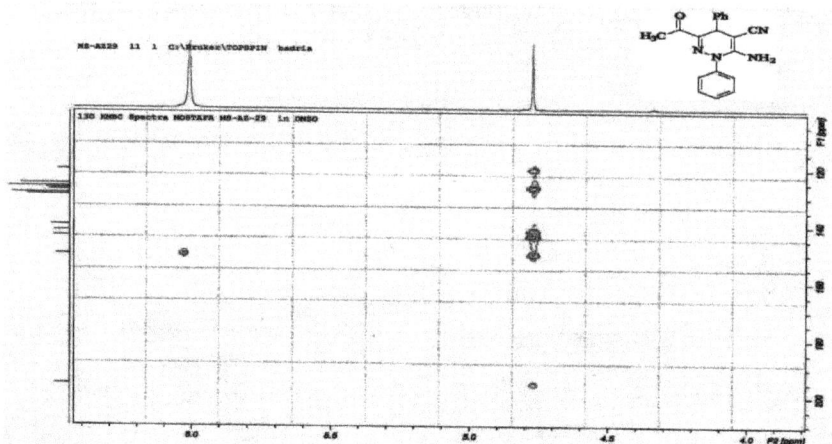

Fig. 46

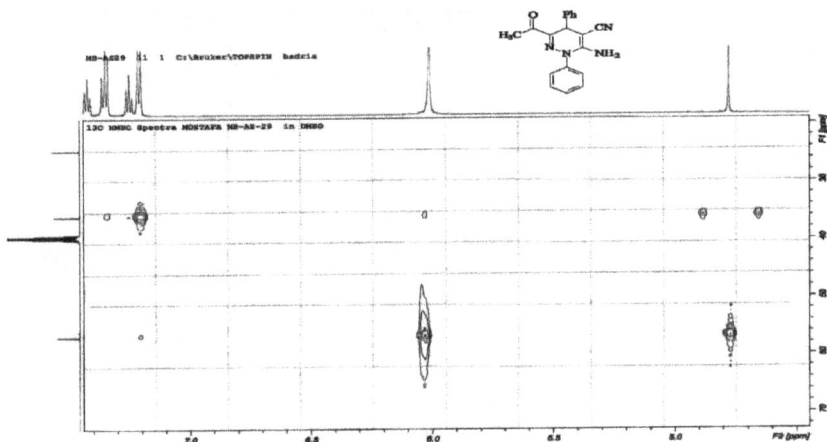

Fig. 47

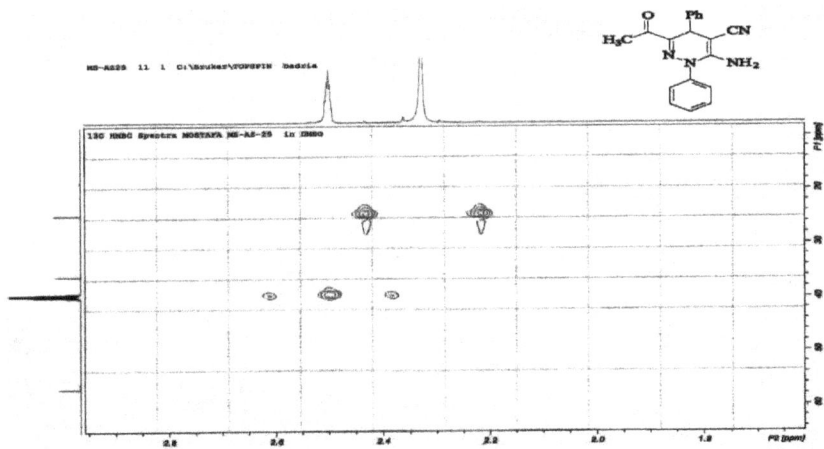

Fig. 48

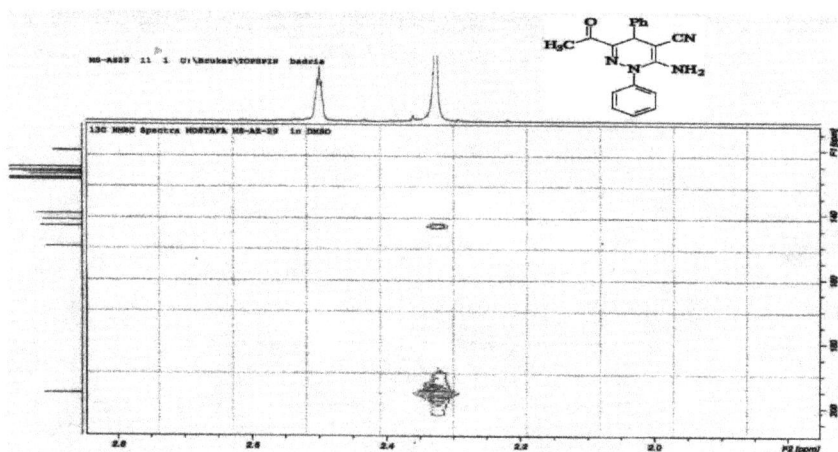

Fig. 49

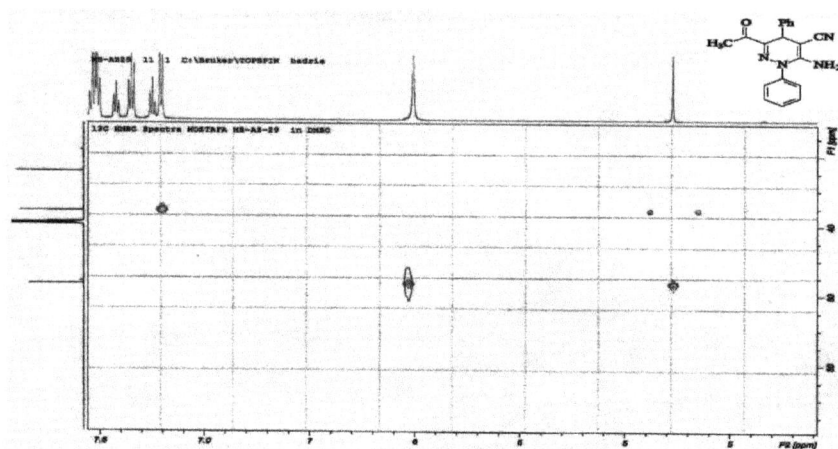

Fig. 50

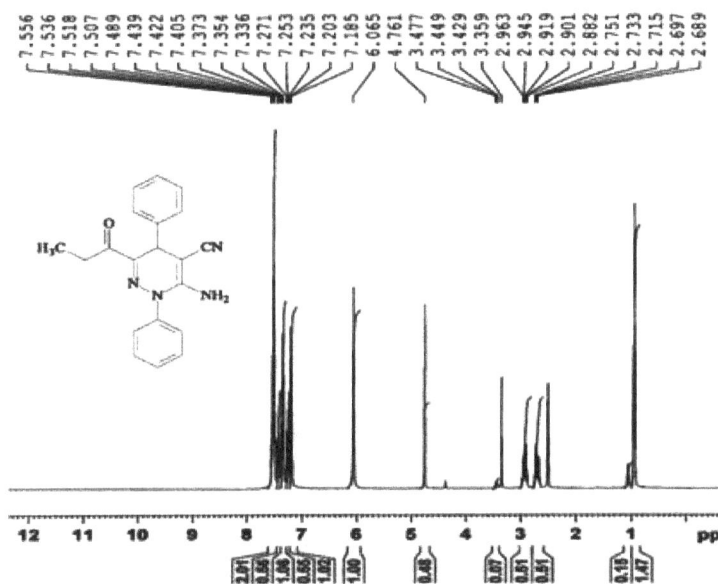

Fig. 51

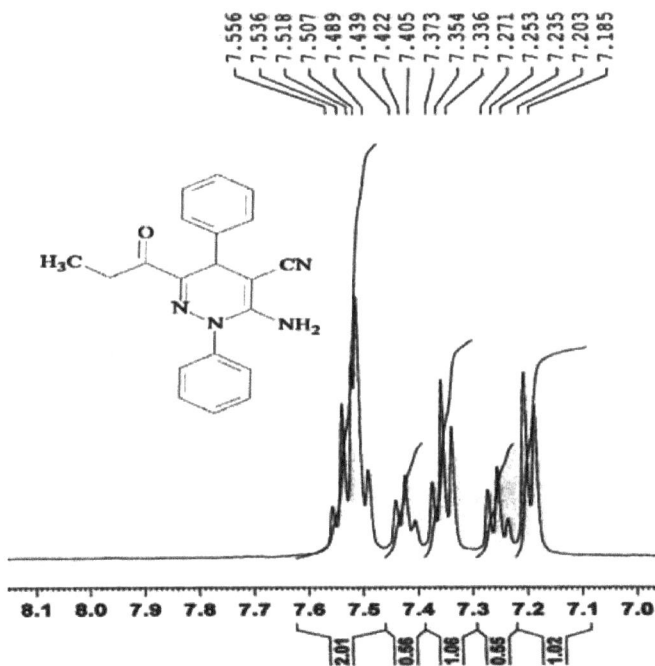

Fig. 52

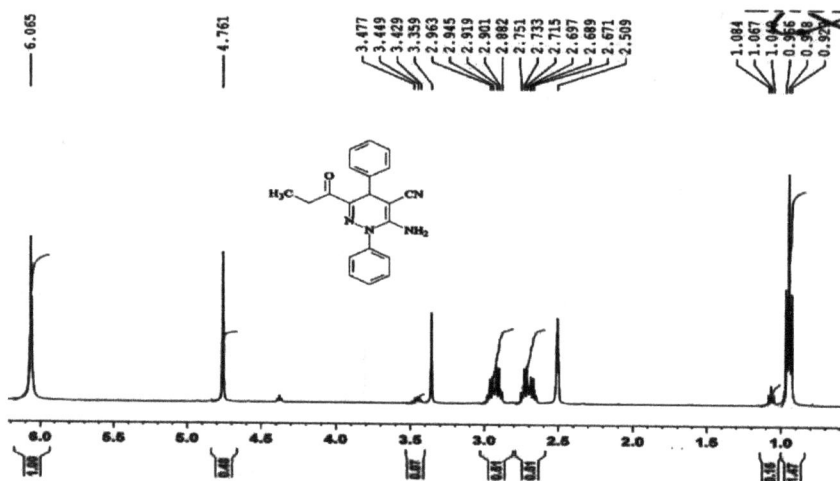

Fig. 53

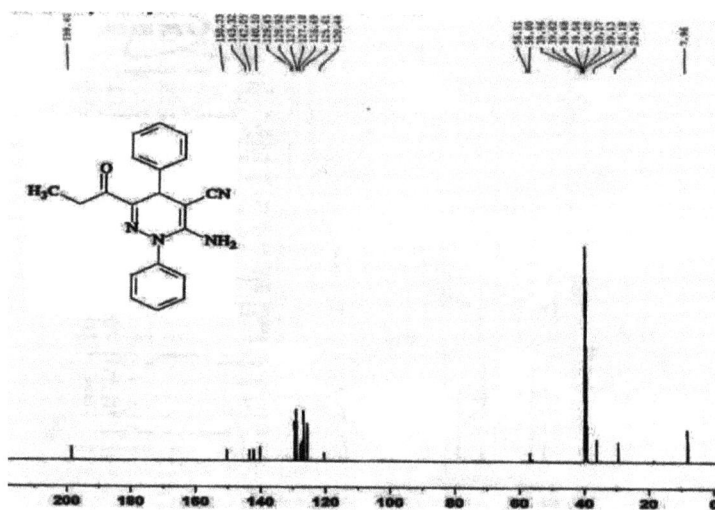

Fig. 54

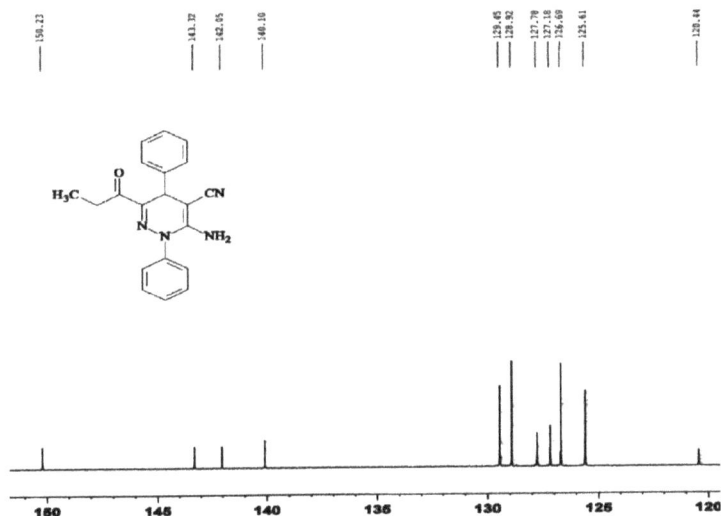

Fig. 55

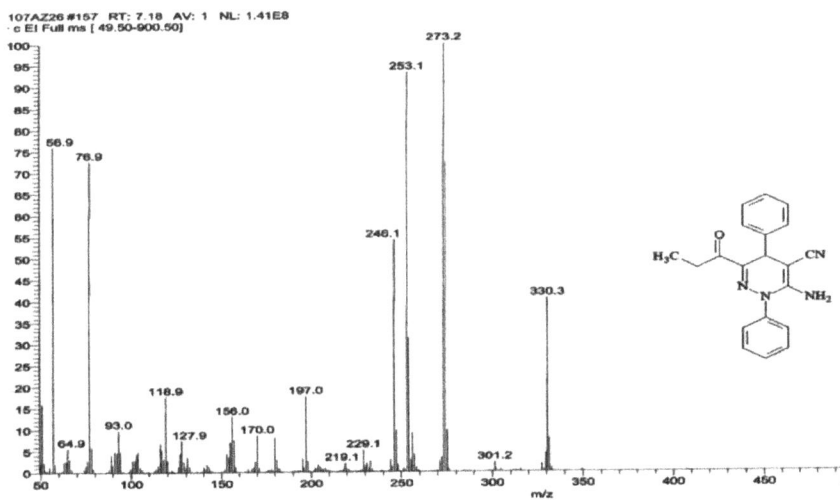

Fig. 56

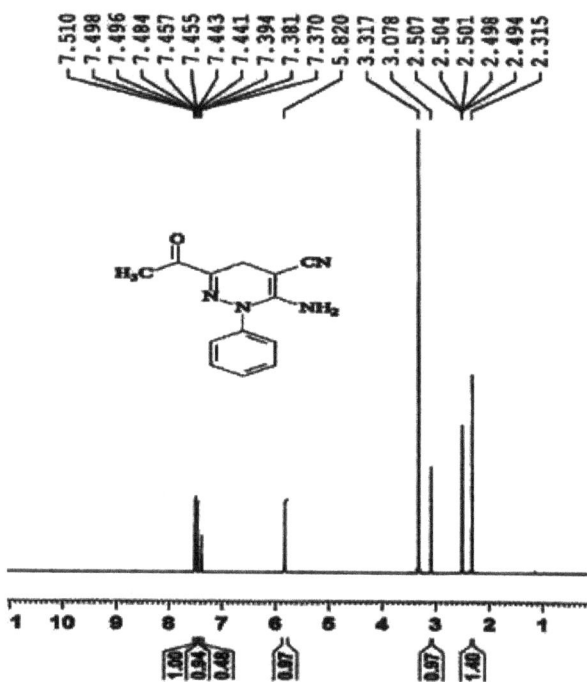

Fig. 57

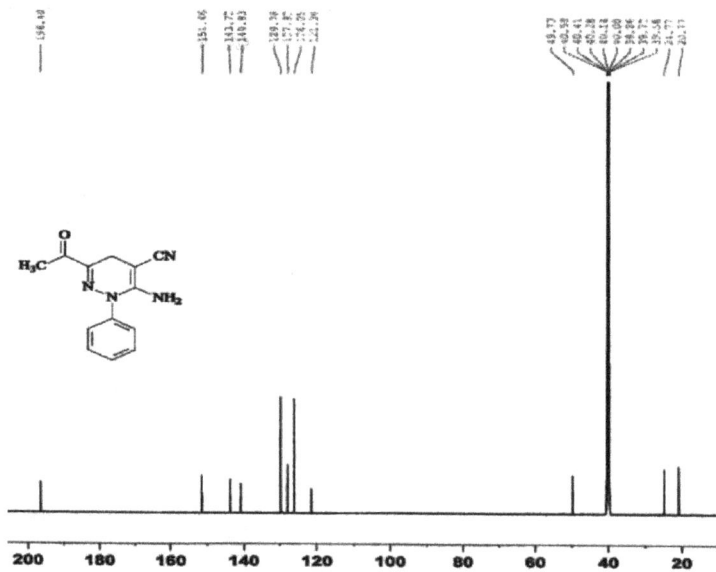

Fig. 58

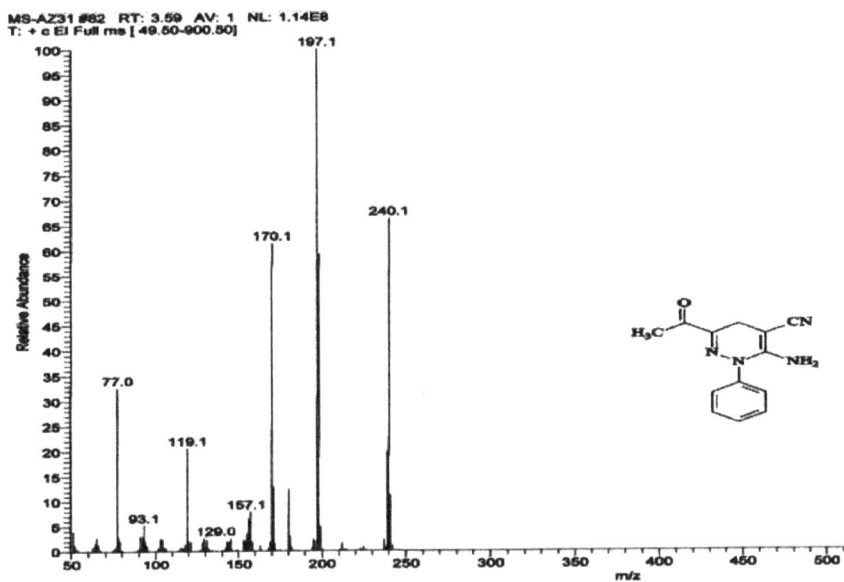

Fig. 59

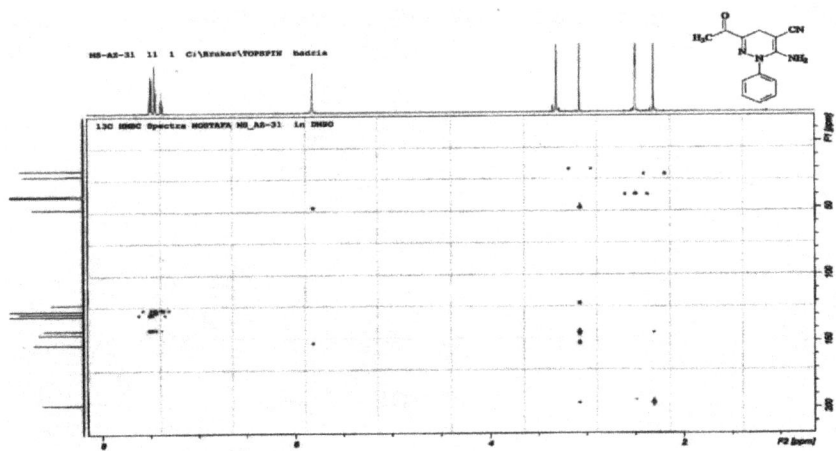

Fig. 60

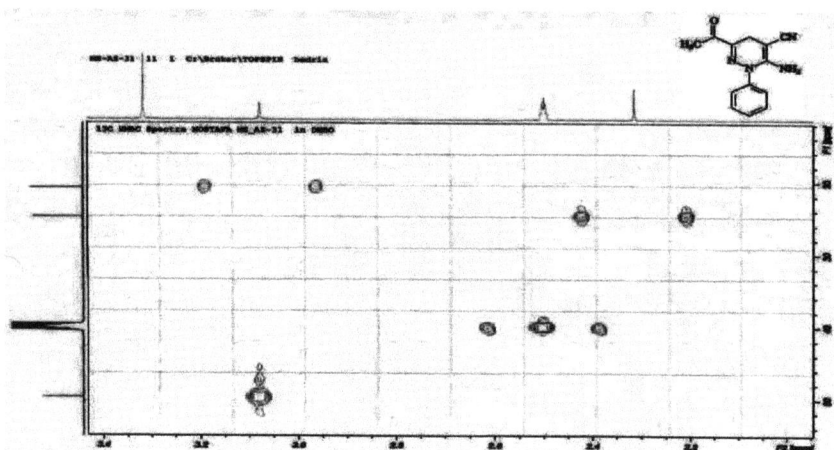

Fig. 61

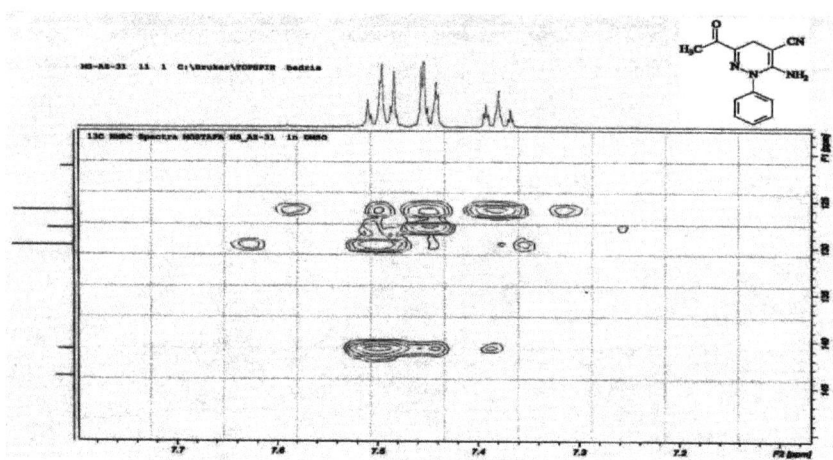

Fig. 62

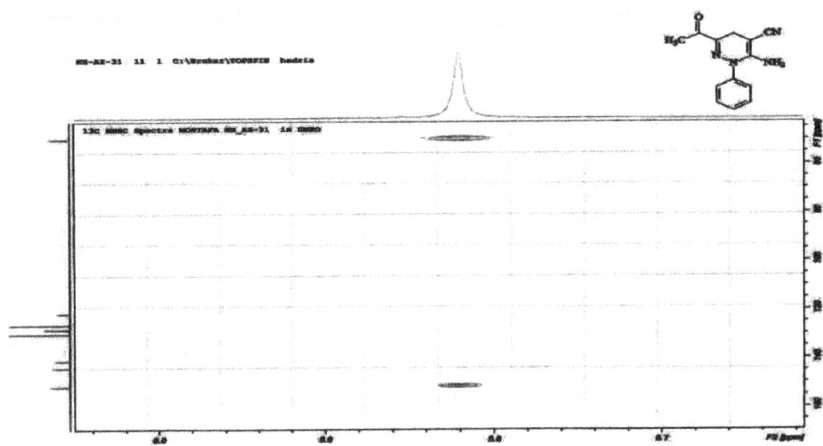

Fig. 63

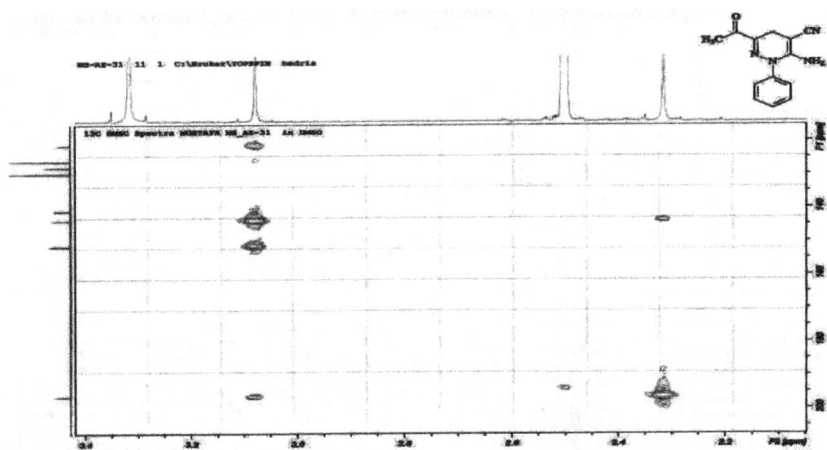

Fig. 64

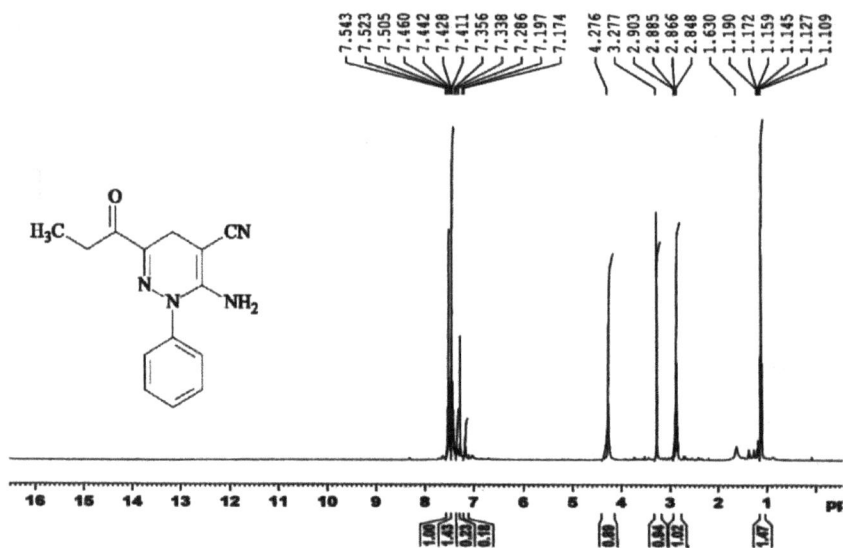

7.543
7.523
7.505
7.460
7.442
7.411
7.356
7.338
7.286
7.197
7.174
4.276
3.277
2.903
2.885
2.866
2.848
1.630
1.190
1.172
1.159
1.145
1.127
1.109

16 15 14 13 12 11 10 9 8 7 6 5 4 3 2 1 ps

Fig. 65

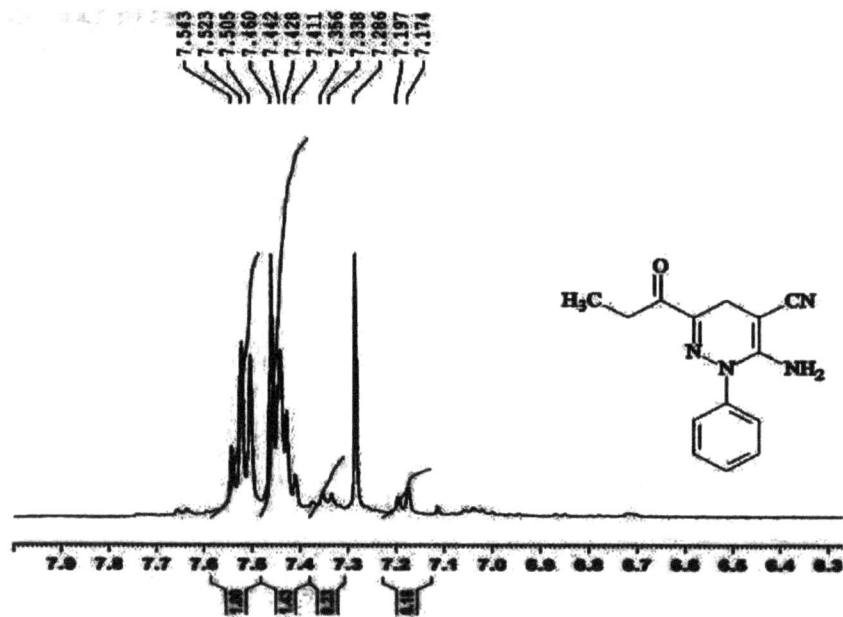

7.543
7.523
7.505
7.460
7.442
7.411
7.356
7.338
7.286
7.197
7.174

7.9 7.8 7.7 7.6 7.5 7.4 7.3 7.2 7.1 7.0 6.9 6.8 6.7 6.6 6.5 6.4 6.3

Fig. 66

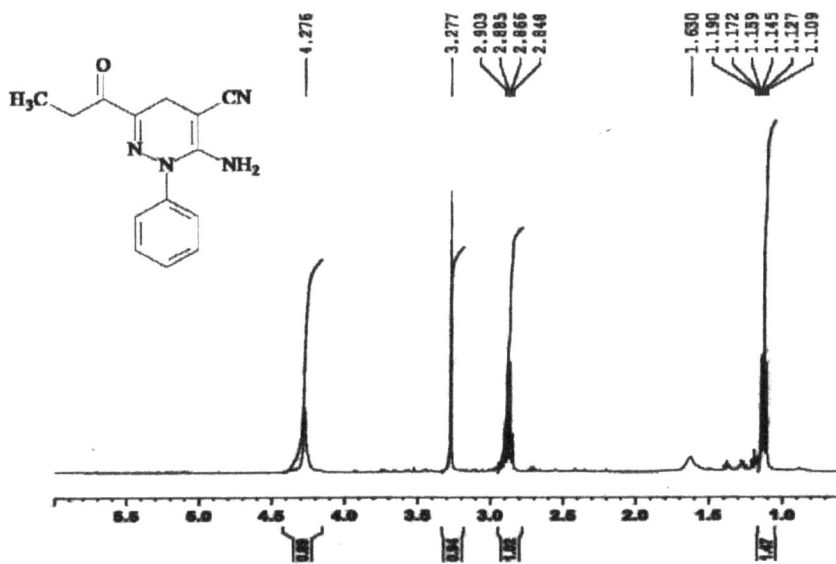

Fig. 67

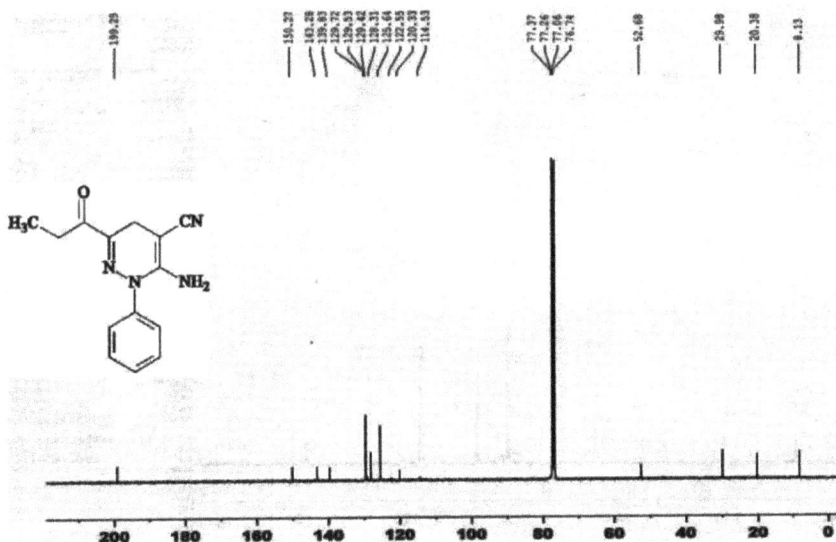

Fig. 68

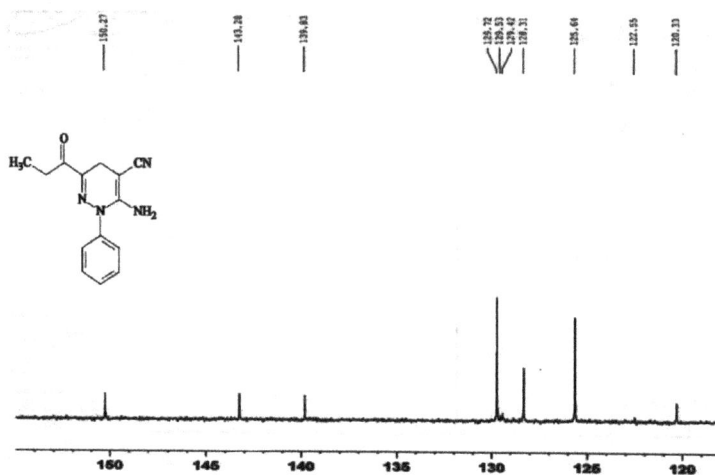

Fig. 69

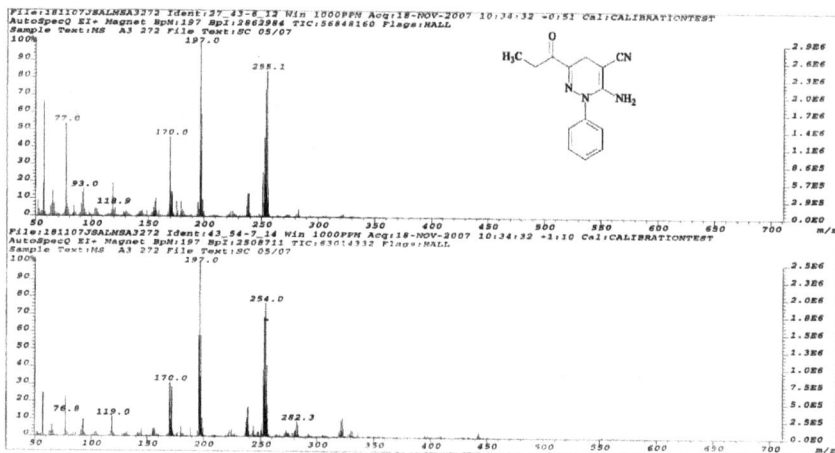

Fig. 70

1H spectra MOSTAFA MS-7 in DMSO

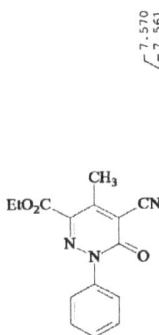

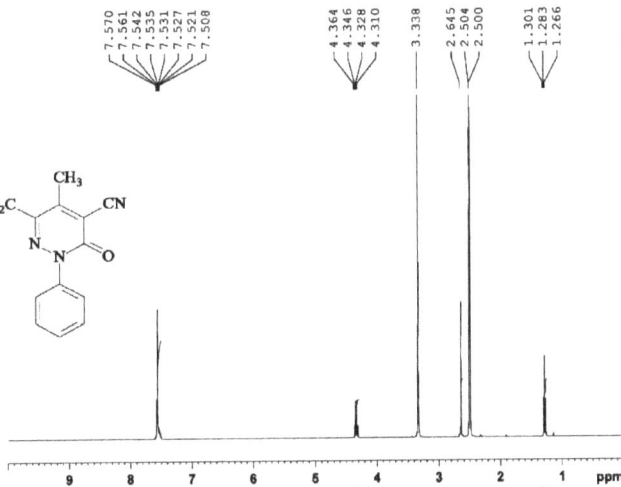

7.570
7.561
7.542
7.535
7.531
7.527
7.521
7.508

4.364
4.346
4.328
4.310

3.338

2.645
2.504
2.500

1.301
1.283
1.266

BRUKER

Current Data Parameters
NAME   Ms-7-1H
EXPNO   1
PROCNO   1

F2 - Acquisition Parameters
Date_   20100330
Time   11.33
INSTRUM   spect
PROBHD   5 mm DUL 13C-1
PULPROG   zg30
TD   65536
SOLVENT   DMSO
NS   8
DS   2
SWH   8278.146 Hz
FIDRES   0.126314 Hz
AQ   3.9584243 sec
RG   574.7
DW   60.400 usec
DE   6.00 usec
TE   673.2 K
D1   1.00000000 sec
TDO   1

========= CHANNEL f1 =========
NUC1   1H
P1   9.00 usec
PL1   -4.50 dB
SFO1   400.1324710 MHz

F2 - Processing parameters
SI   32768
SF   400.1300028 MHz
WDW   EM
SSB   0
LB   0.30 Hz
GB   0
PC   1.00

2.53   1.00   1.51   1.50

Fig. 71

13C decoupled spectra Dr. Saleh MS-7 in DMSO

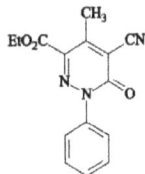

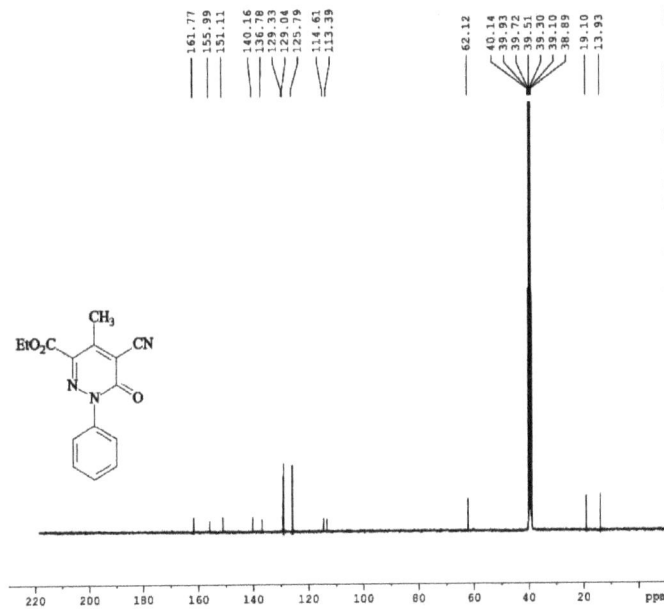

161.77
155.99
151.11

140.16
136.78
129.04
129.33
125.79

114.61
113.39

62.12

40.14
39.93
39.72
39.51
39.30
39.10
38.89

19.10
13.93

BRUKER

Current Data Parameters
NAME   MS-7-13C
EXPNO   2
PROCNO   1

F2 - Acquisition Parameters
Date_   20100401
Time   0.55
INSTRUM   spect
PROBHD   5 mm DUL 13C-1
PULPROG   zgpg30
TD   10240
SOLVENT   DMSO
NS   15360
DS   4
SWH   23980.814 Hz
FIDRES   2.341877 Hz
AQ   0.2135540 sec
RG   1448.2
DW   20.850 usec
DE   6.00 usec
TE   673.2 K
D1   2.00000000 sec
d11   0.03000000 sec
DELTA   1.99999998 sec
TDO   1

========= CHANNEL f1 =========
NUC1   13C
P1   7.20 usec
PL1   -6.00 dB
SFO1   100.6228298 MHz

========= CHANNEL f2 =========
CPDPRG2   waltz16
NUC2   1H
PCPD2   80.00 usec
PL2   -3.00 dB
PL12   15.06 dB
PL13   18.00 dB
SFO2   400.1316005 MHz

F2 - Processing parameters
SI   32768
SF   100.6126155 MHz
WDW   EM
SSB   0
LB   1.00 Hz
GB   0
PC   1.40

Fig. 72

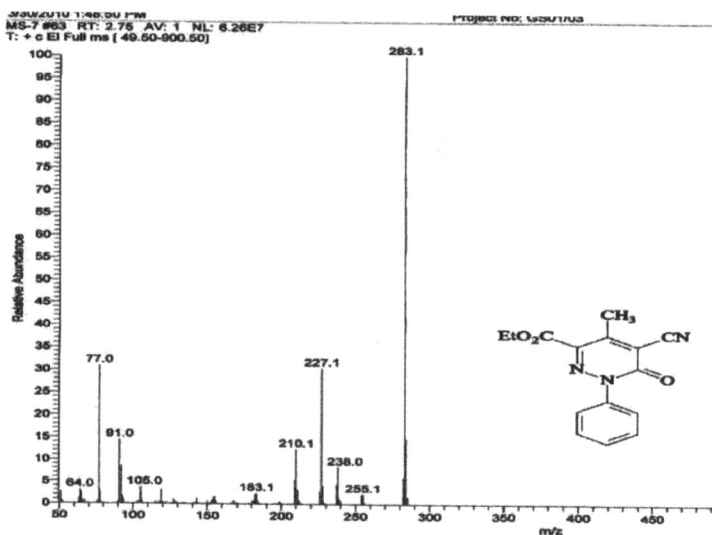

Fig. 73

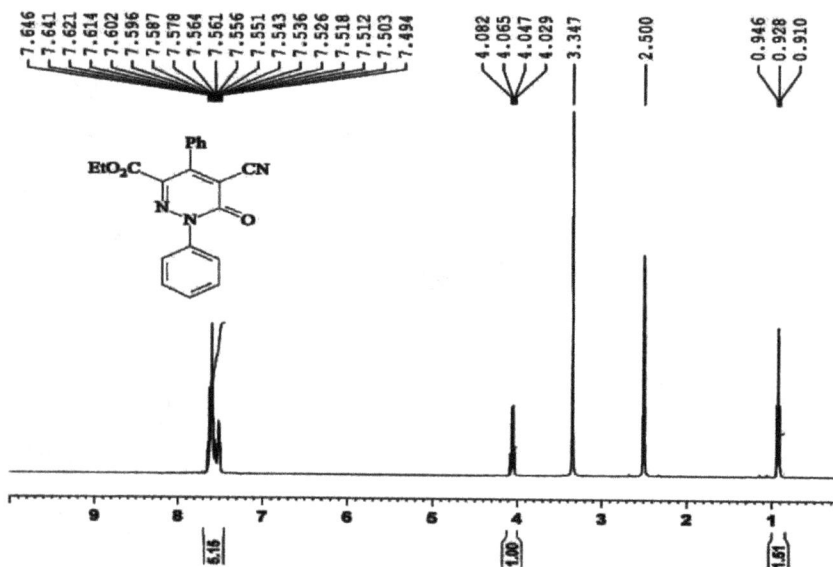

Fig. 74

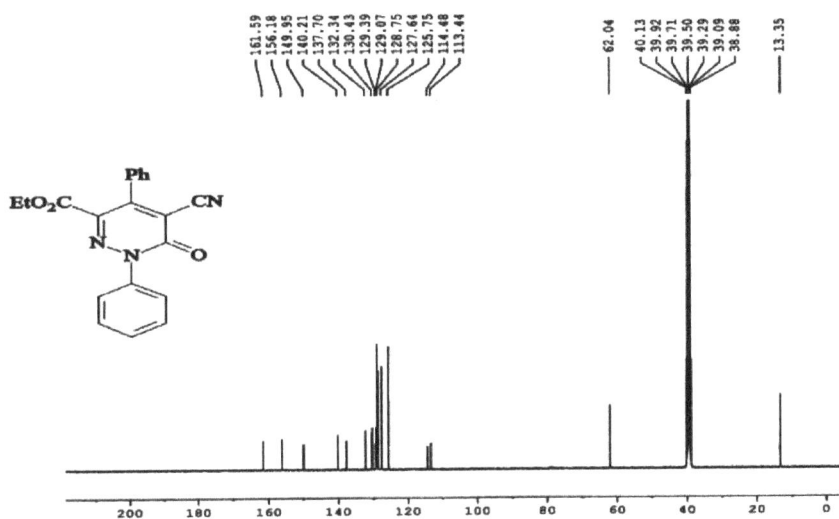

Fig. 75

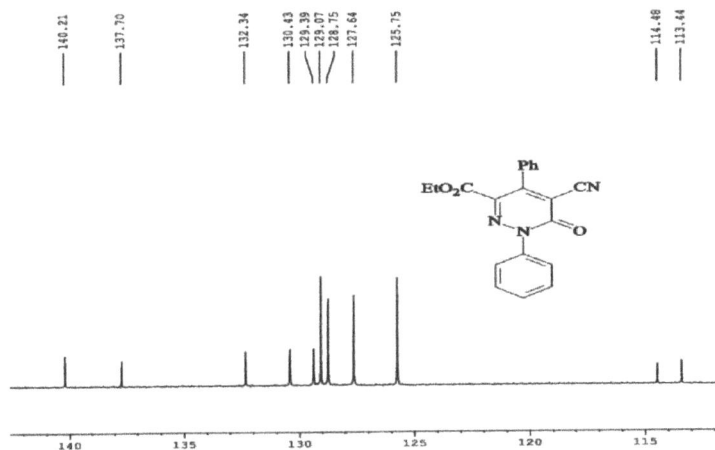

Fig. 76

Problem 4: In 1999, we reported the reaction given in Eq. (7). However when we recently reinspected the spectral (Figs. 78–83) data, we concluded that the product was actually a result of the reaction pathway given in Eq. (8). Discuss why and whether it was possible to distinguish a

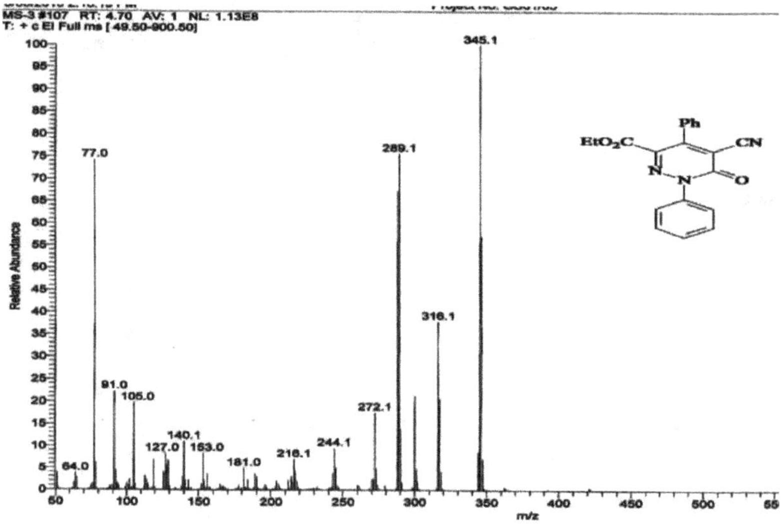

Fig. 77

possibly formed pyranimine from the formed nicotinates structure. Give examples rearrangement is shown in Eq. (8).

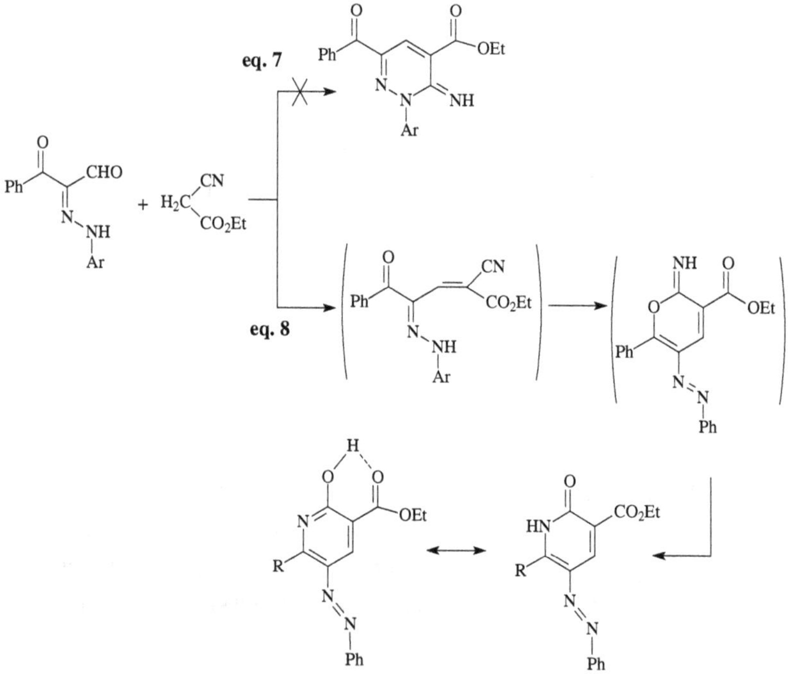

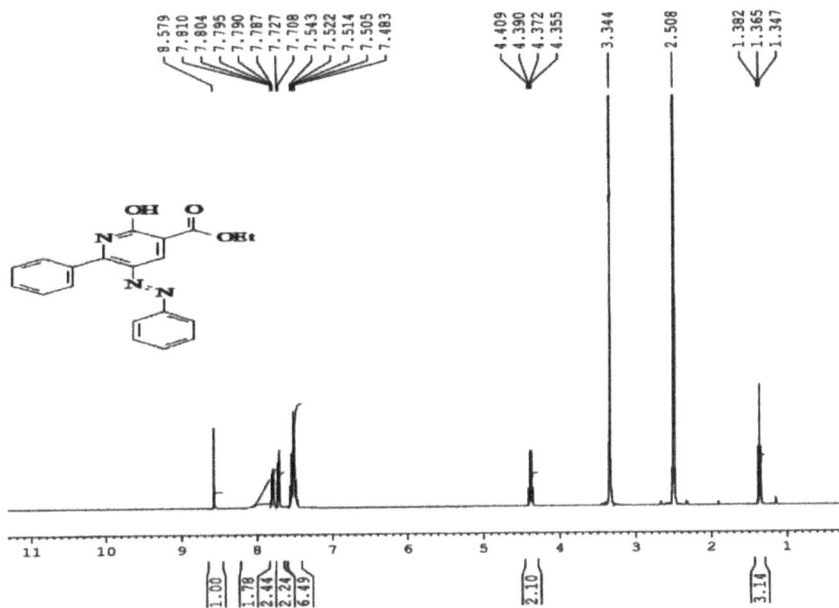

Fig. 78

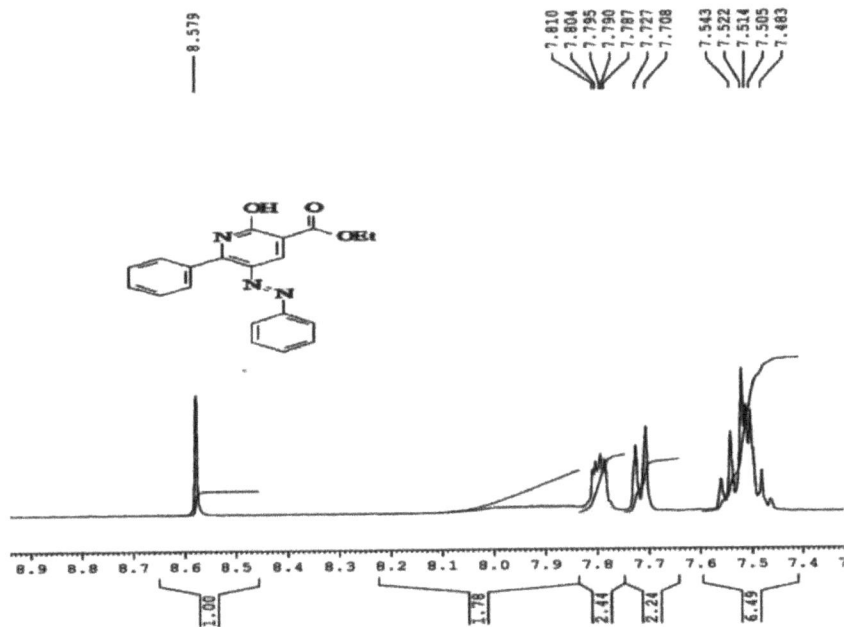

Fig. 79

Spectroscopic Identification of Organic Molecules

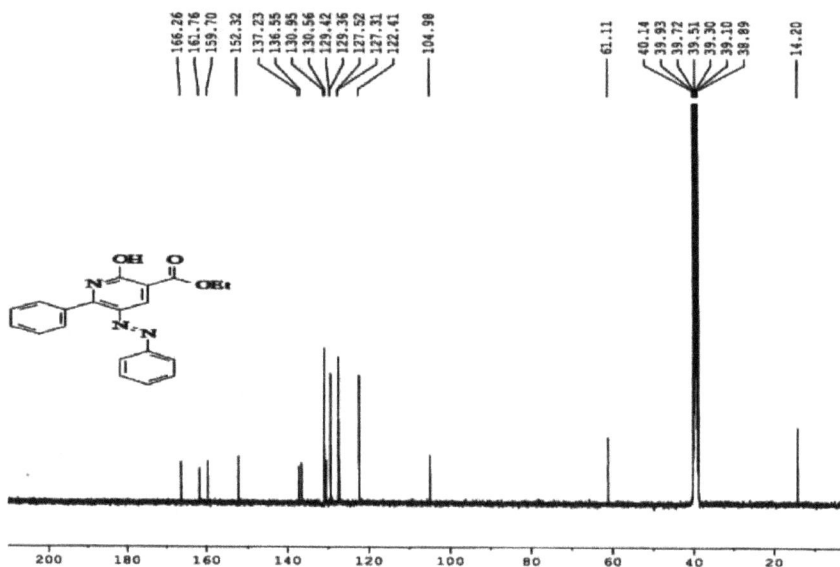

Fig. 80

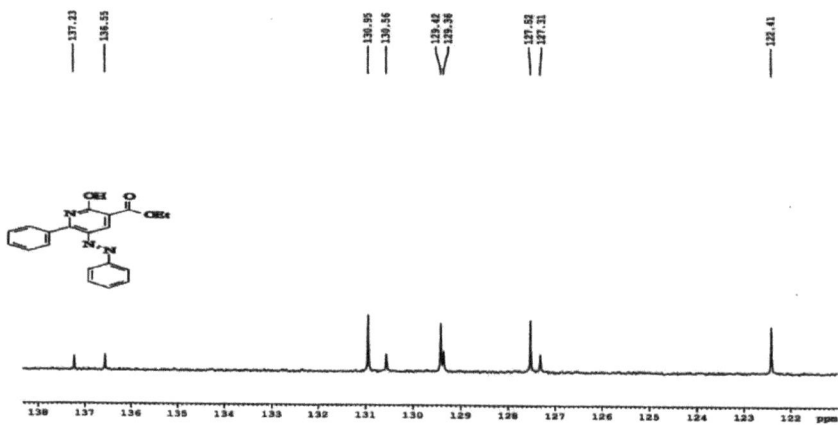

Fig. 81

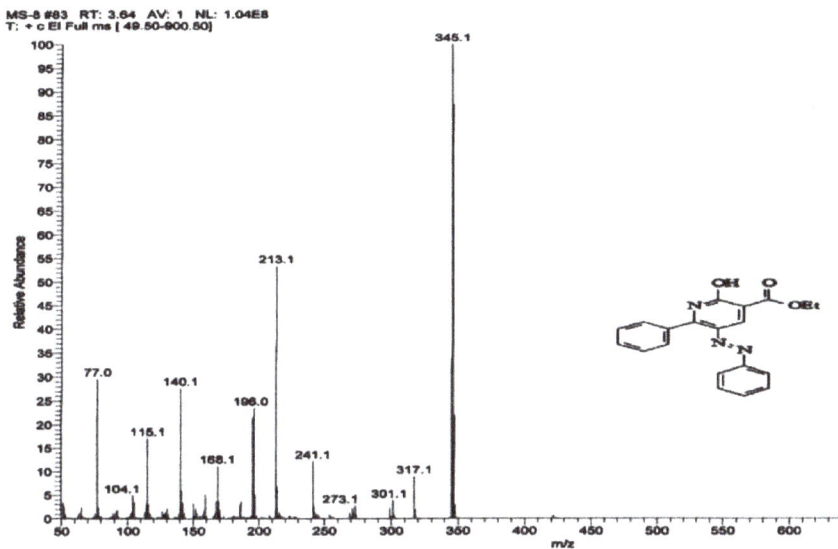

Fig. 82

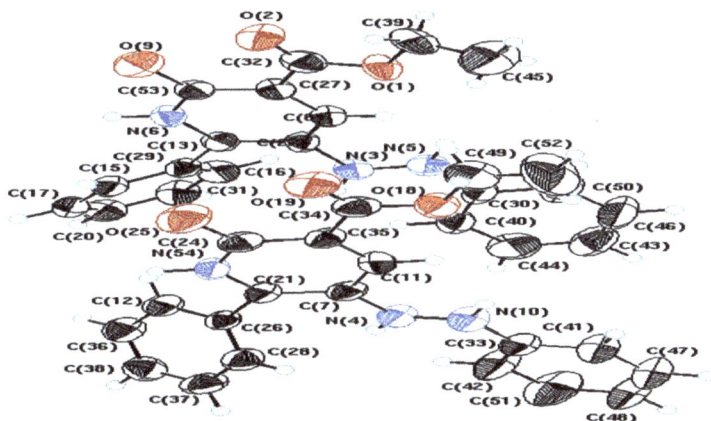

Fig. 83

Problem 5: Some time ago, we were reacted arylhydrazonals with hippuric acid in the presence of Ac_2O. A product was obtained with molecular formula and preliminary data that seemed at first glance interpretable for both benzoylaminopyridazine and arylazobenzoylaminopyrans. Careful inspection of the data led to the conclusion of the structure; show which one and why (Figs. 84–88).

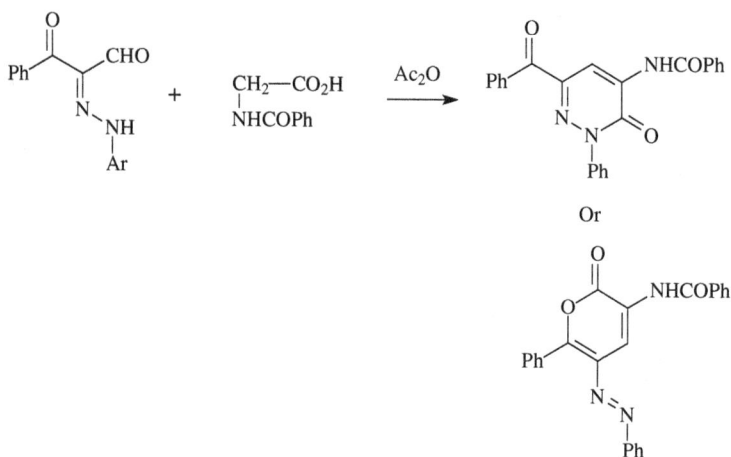

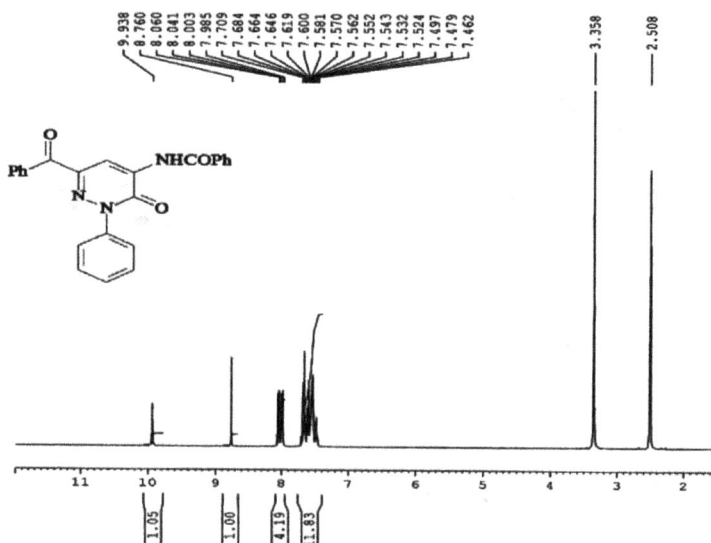

Fig. 84

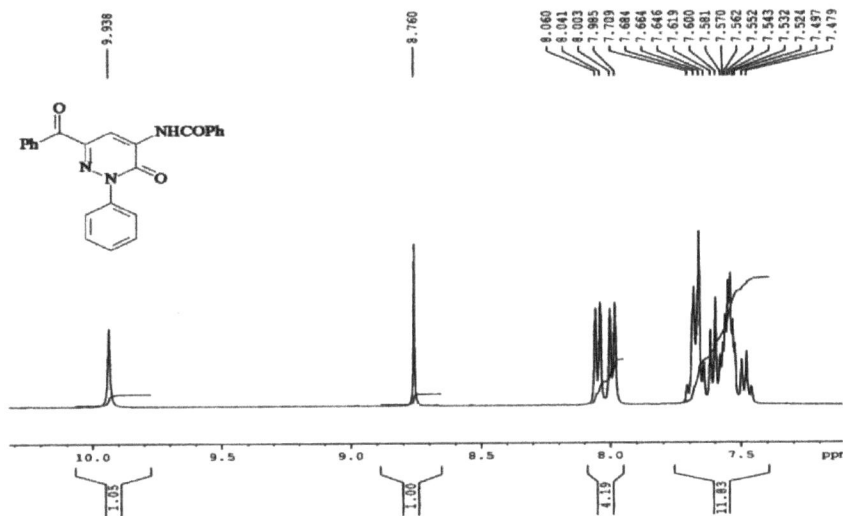

Fig. 85

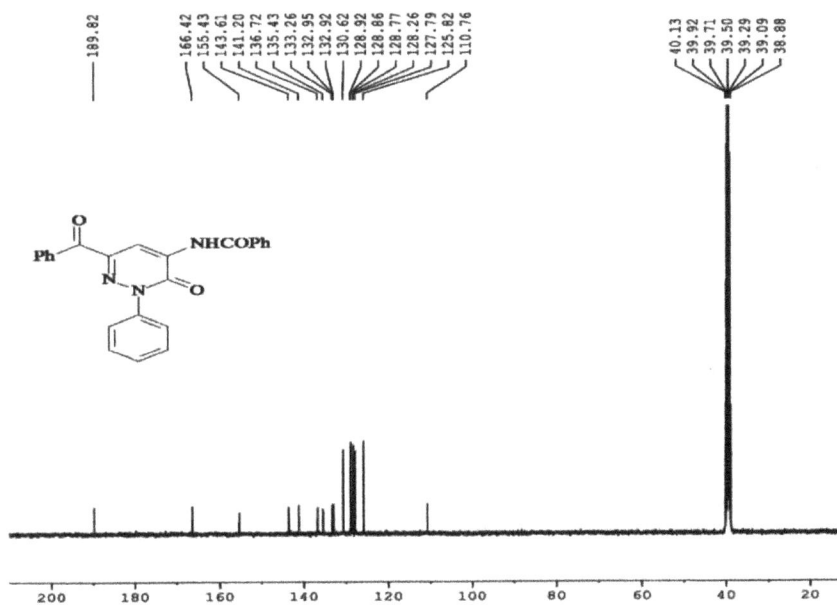

Fig. 86

143.61

141.20

136.72

135.43

133.26
132.95
132.92

130.62

128.92
128.86
128.77
128.26

127.79

125.82

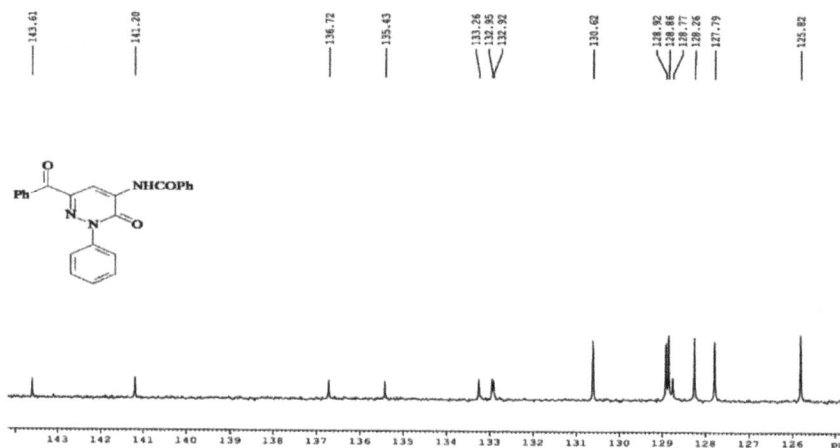

Fig. 87

ms-2 #136 RT: 5.99 AV: 1 NL: 7.31E7
T: + c EI Full ms [49.50-900.50]

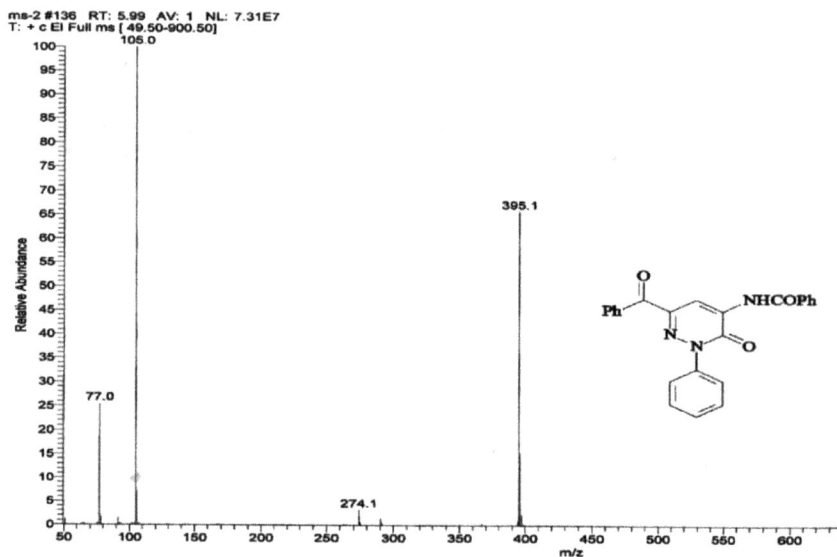

Fig. 88

Problem 6: The data in Figs. 89–93 were obtained for the product obtained on reacting phenylhydrazonals with dimethylacetylenedicarboxylate. These were interpreted for the pyridazine structure shown in Eq. (9). When the work was published, one of the referees asked for evidence against the phenylazopyran structure. This question could be readily and easily answered from the data. Show how to interpret ^{1}H NMR signals and the most significant ^{13}C NMR signals.

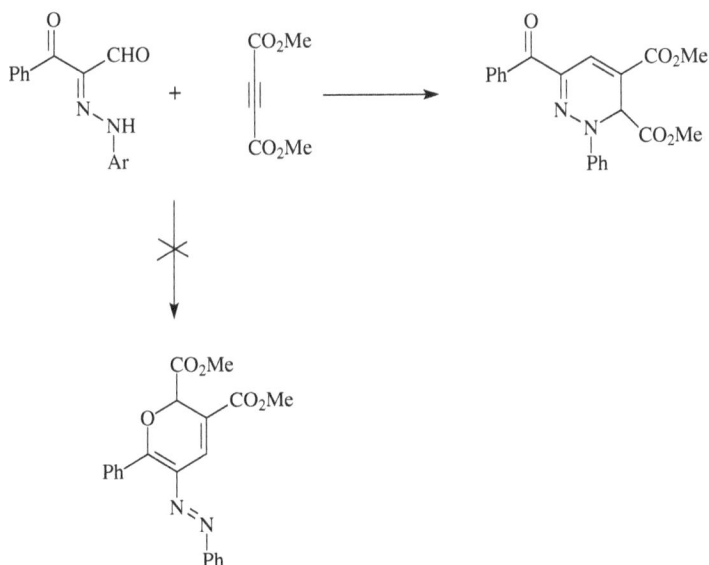

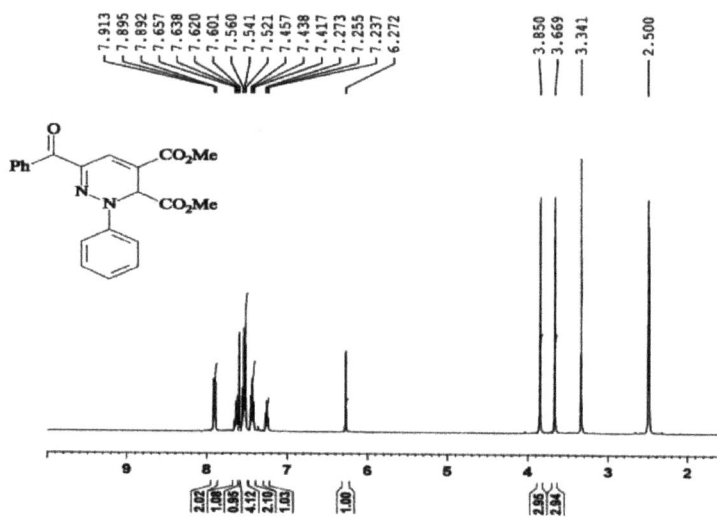

Fig. 89

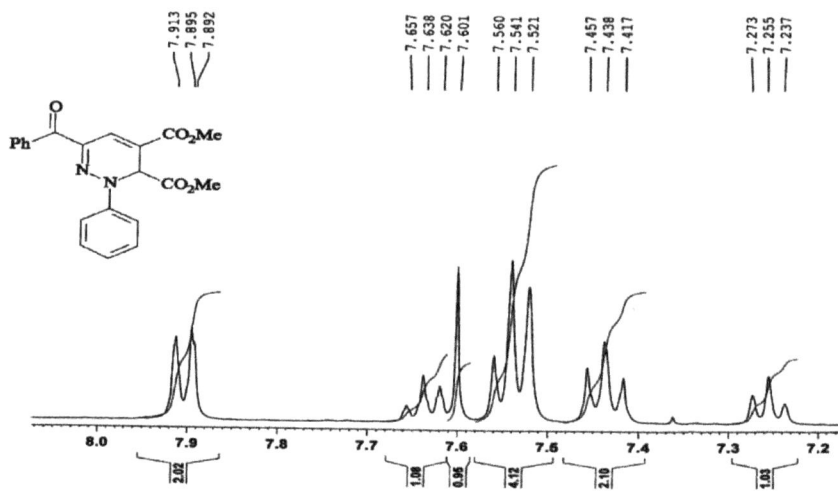

Fig. 90

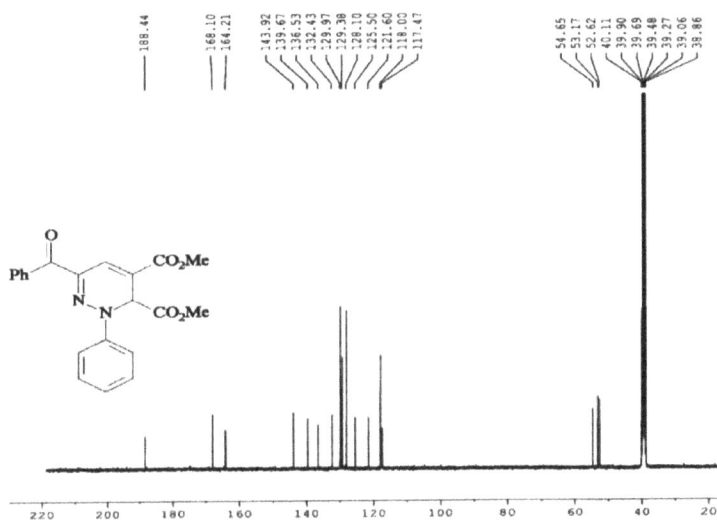

Fig. 91

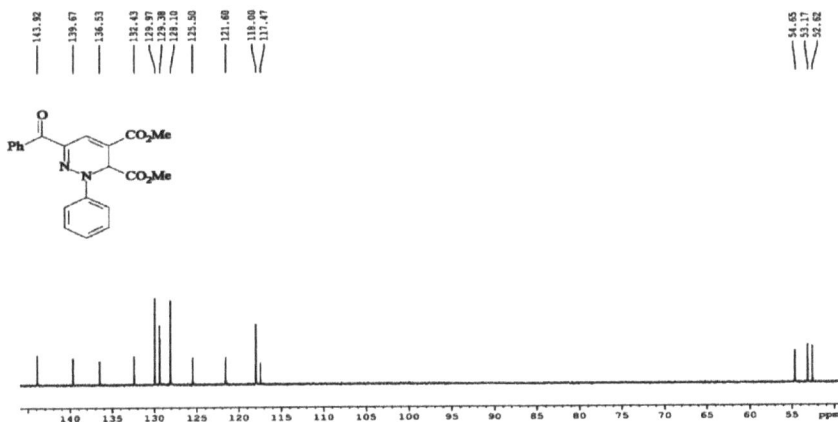

Fig. 92

Problem 7: The phenylhydrazonal (**A**) was treated with acrylonitrile in the presence of DABCO, and a product of condensation via water elimination was formed. It is assumed that initially the Baylus–Hillman adduct (**B**) was formed. This was assigned structure (**C**); inspect the data and explain why structure (**D**) was ruled out. Also, calculate the J value for the signal at $\delta = 4.8$ ppm and elucidate what the signal at $\delta = 7.88$ stands for (Figs. 94–101).

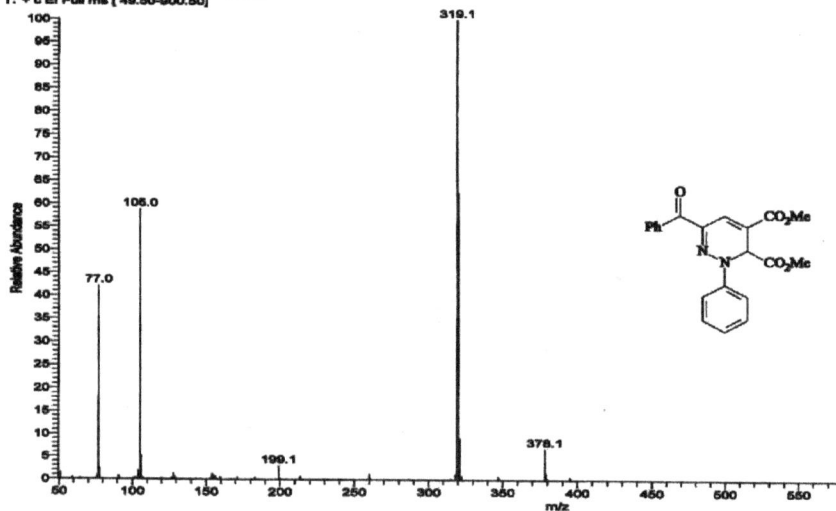

Fig. 93

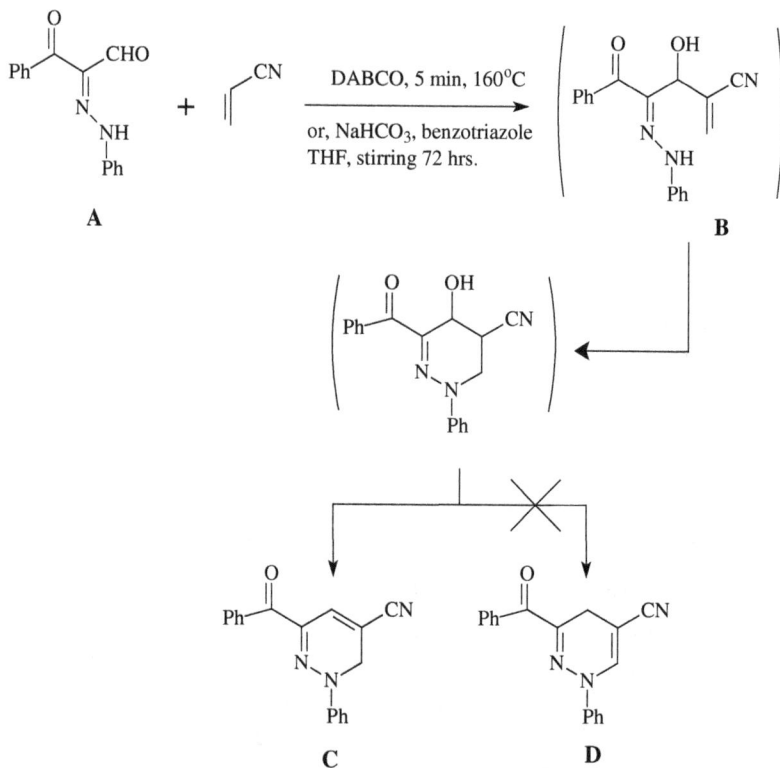

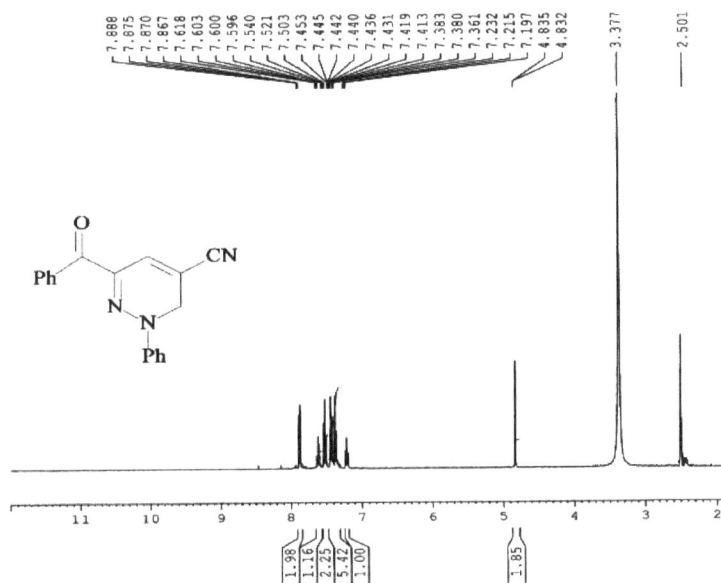

Fig. 94

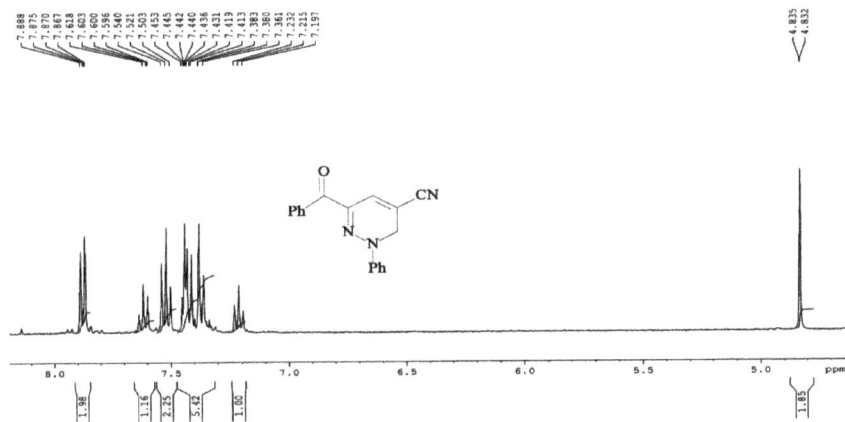

Fig. 95

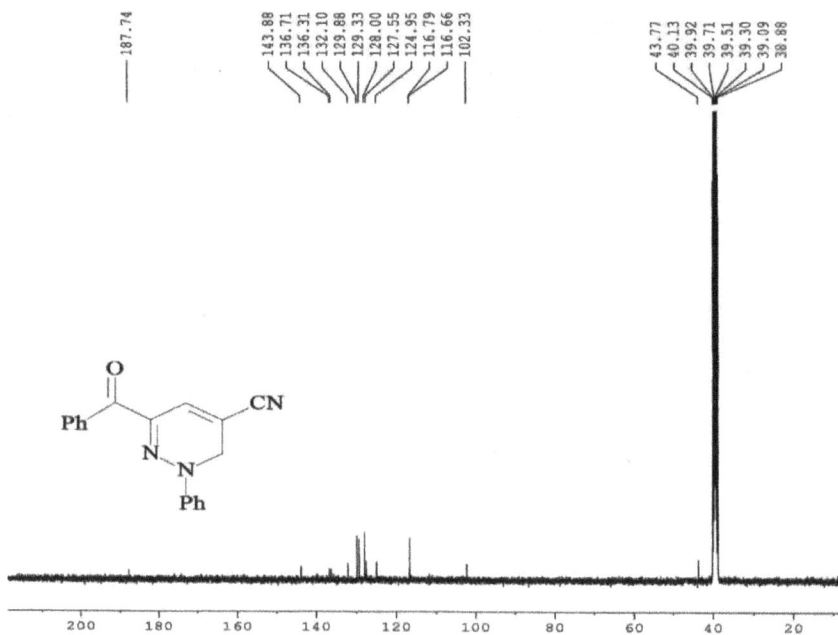

Fig. 96

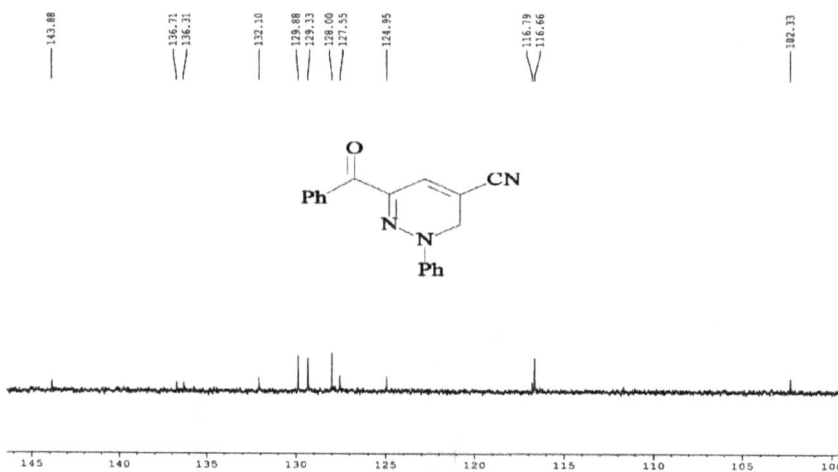

Fig. 97

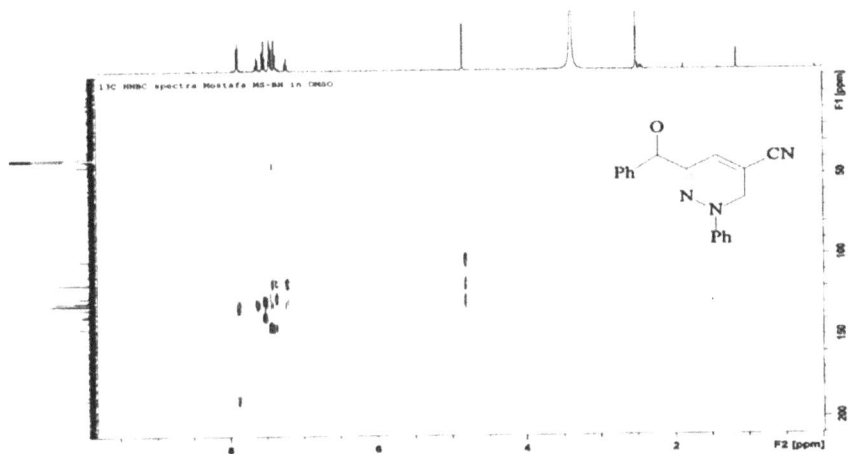

Fig. 98

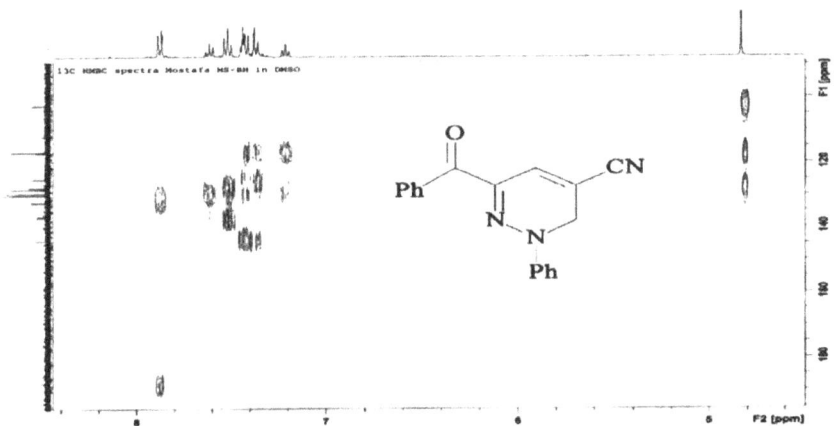

Fig. 99

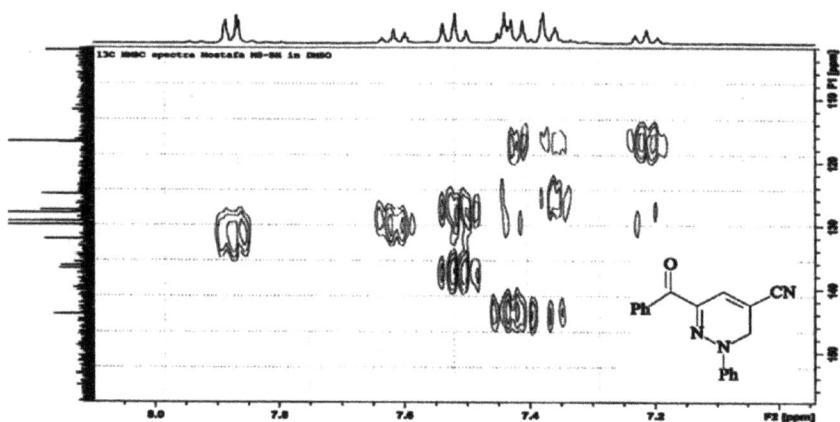

Fig. 100

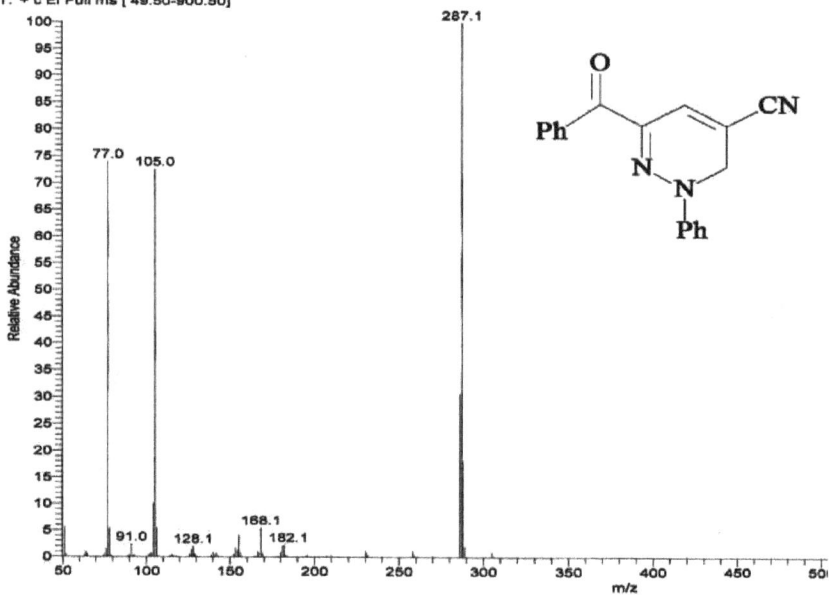

Fig. 101

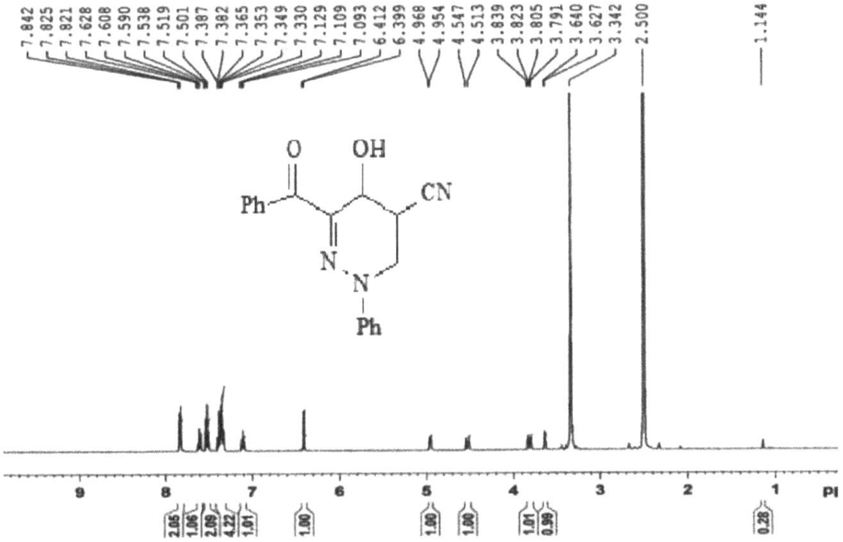

Fig. 102

Problem 8: From the previous reaction, an intermediate adduct could be isolated. It could either be a cyclic adduct (**E**) or pyridazine (**F**). Inspect data in Figs. 102–107 to arrive at the conclusion.

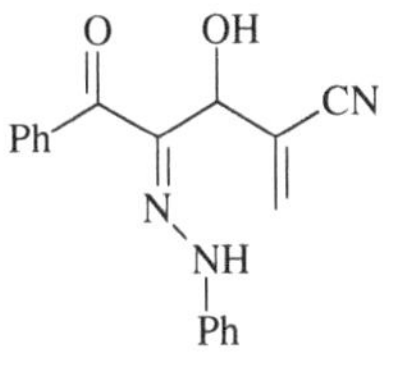

E   F

Or

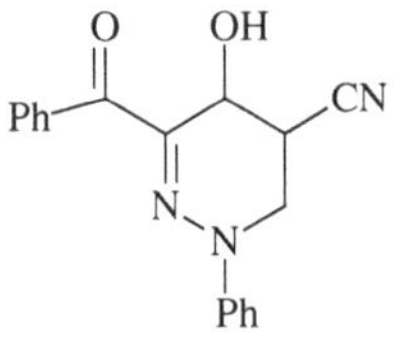

Problem 9: In Figs. 108–111, the ^{1}H NMR and ^{13}C NMR data for phenacylmalononitrile are reported. Indicate the most significant signals that can stand against potential aminofuran structure for this product.

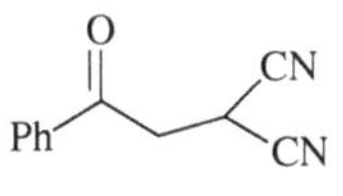

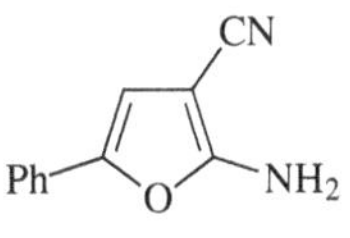

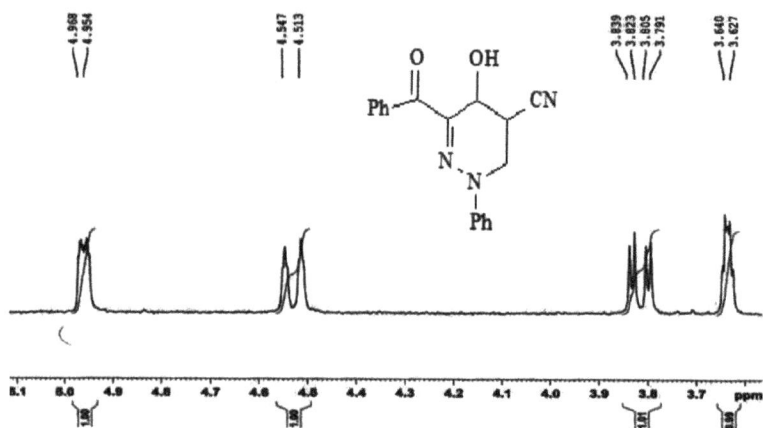

Fig. 103

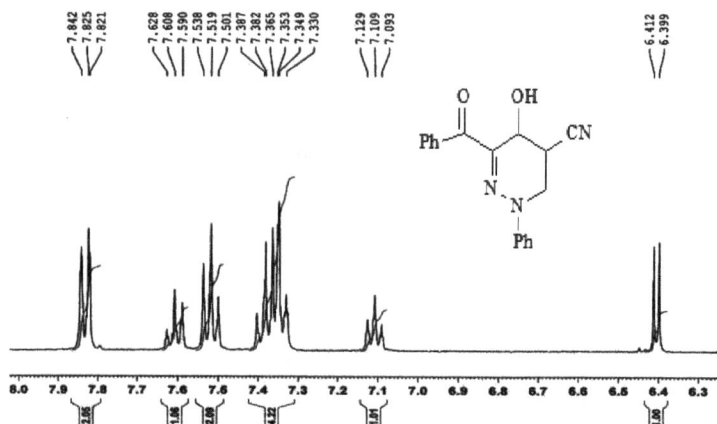

Fig. 104

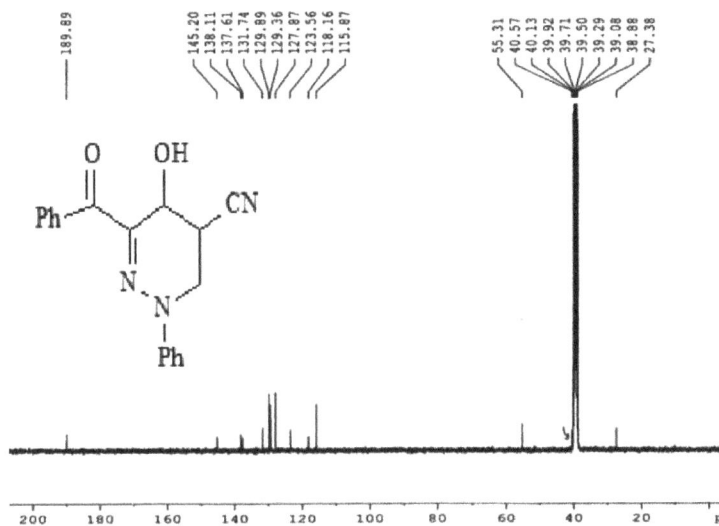

Fig. 105

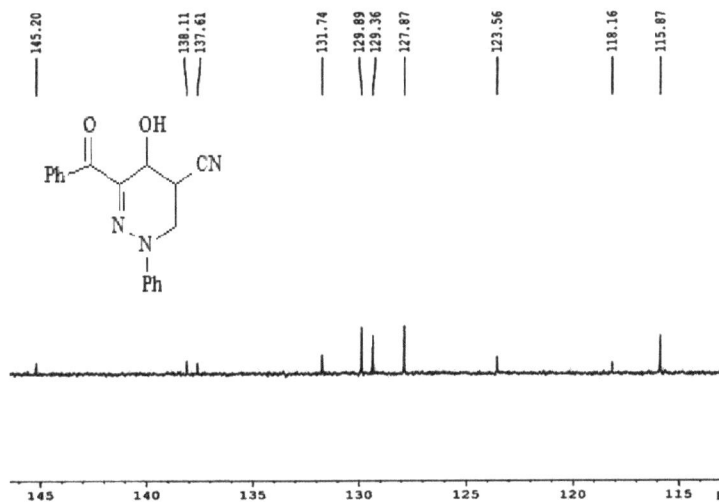

Fig. 106

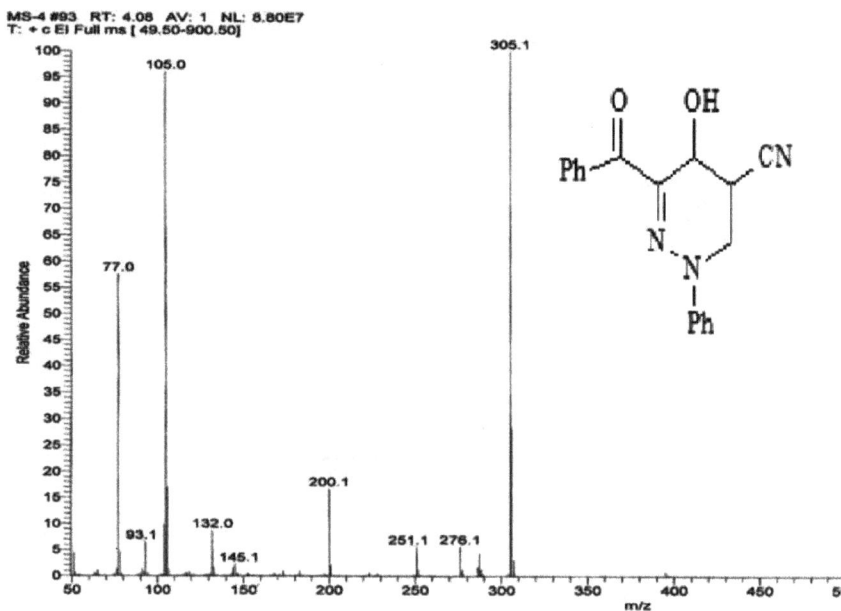

Fig. 107

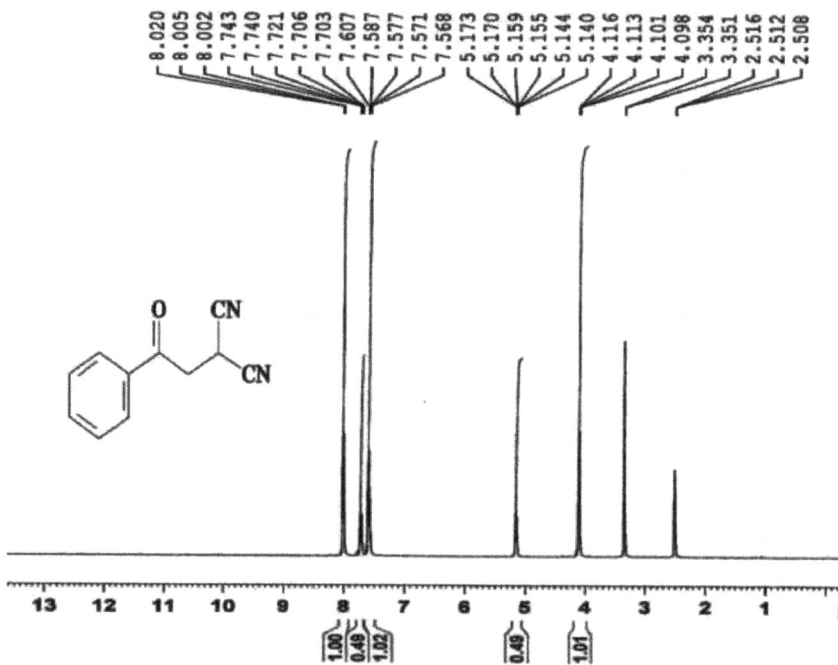

Fig. 108

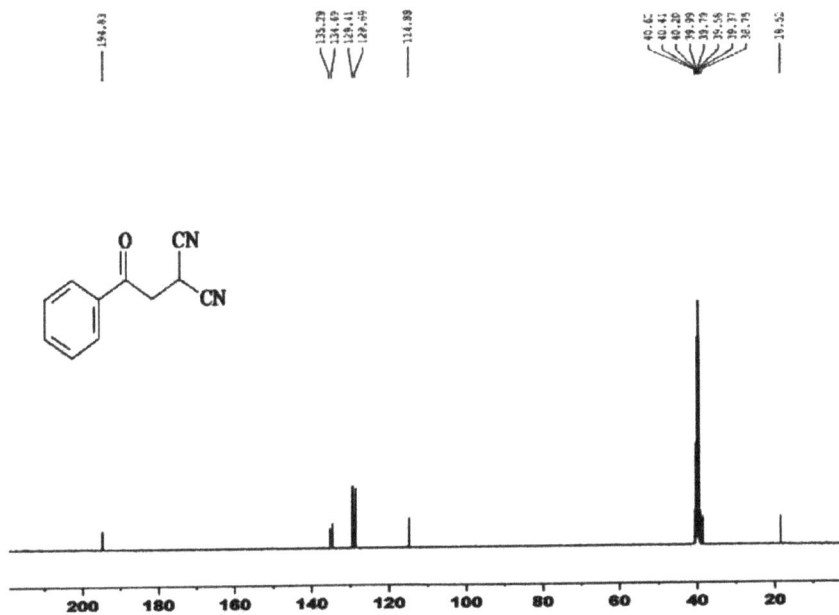

Fig. 109

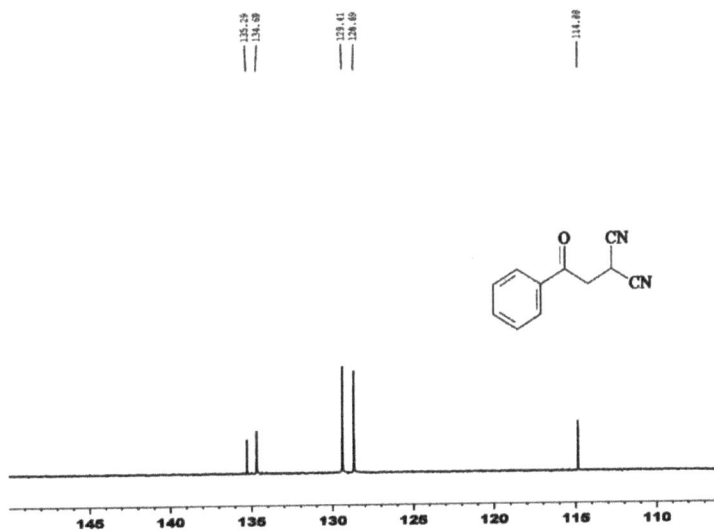

Fig. 110

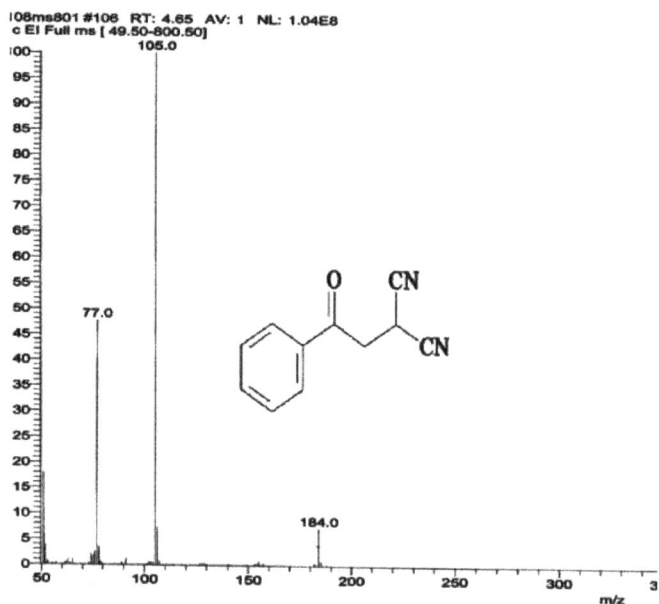

Fig. 111

Problem 11: It has been reported in the literature that phenacylmalononi-trile reacts with hydrazine hydrate to yield phenacylpyrazole-3,5-diamine (see equation below).

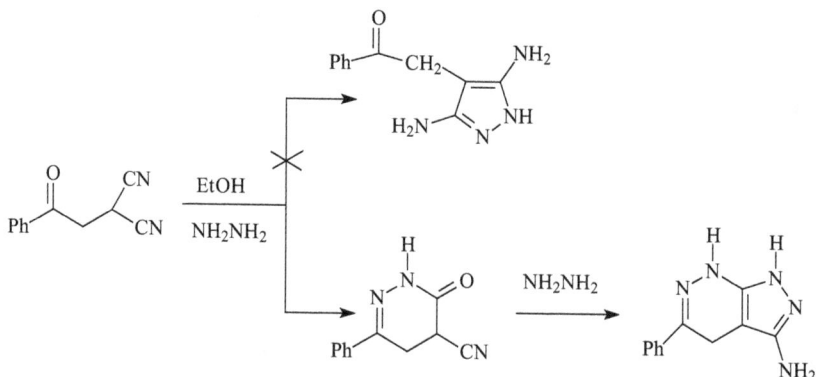

However, repeating this work led to the formation of a product whose spectral data are shown in Figs. 112–118. When this product was heated further with hydrazine hydrate, the product whose spectral data are shown in Figs. 119–121 was formed. Use these data to deduce the structure of both the products.

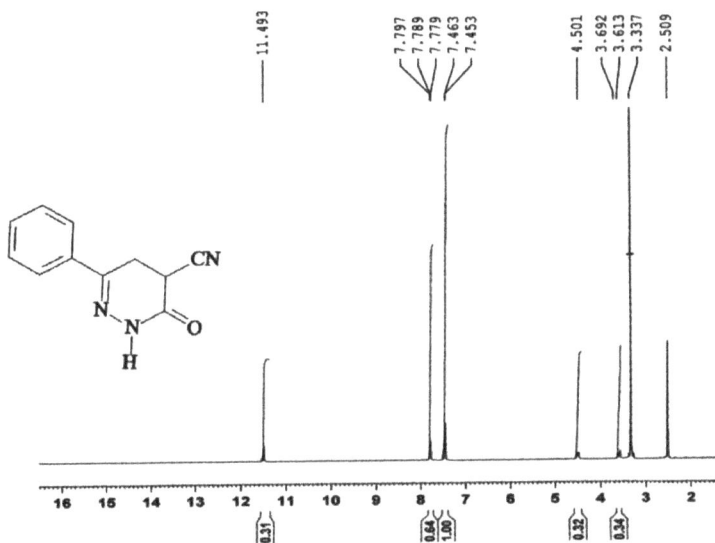

Fig. 112

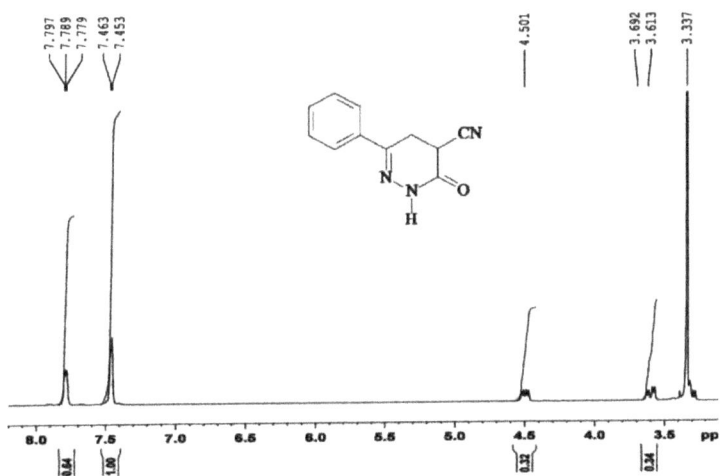

Fig. 113

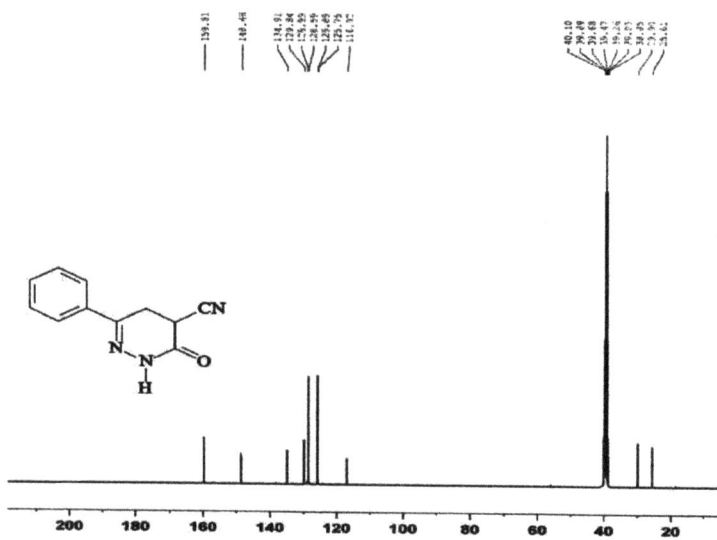

Fig. 114

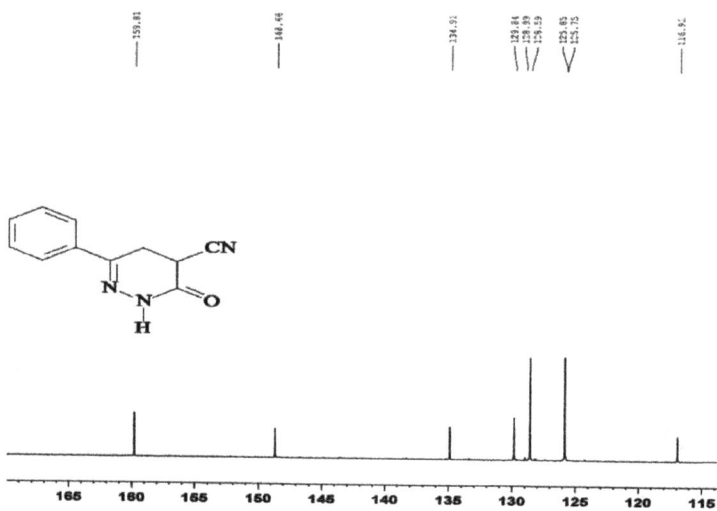

Fig. 115

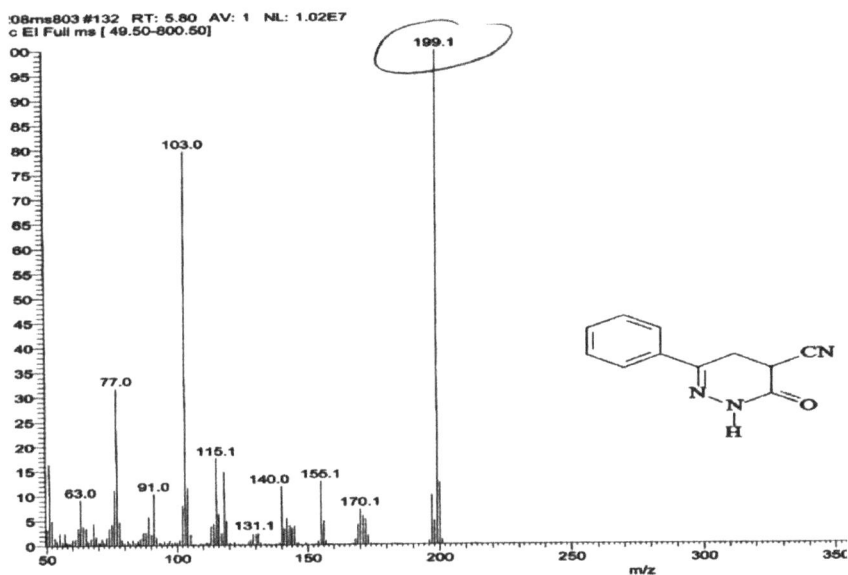

Fig. 116

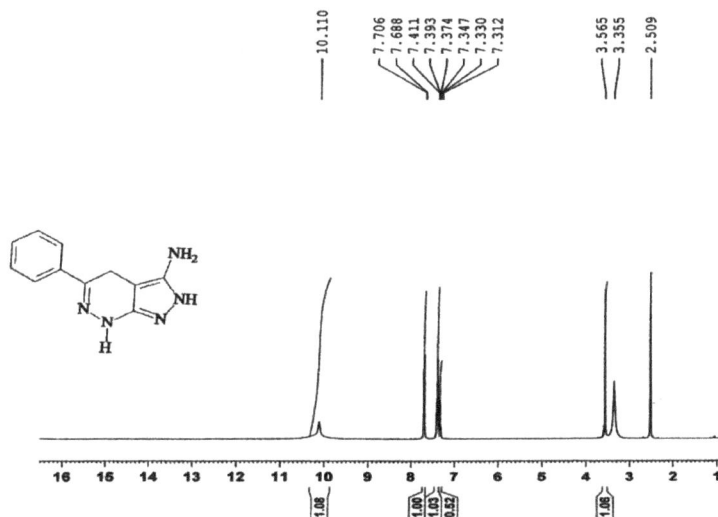

Fig. 117

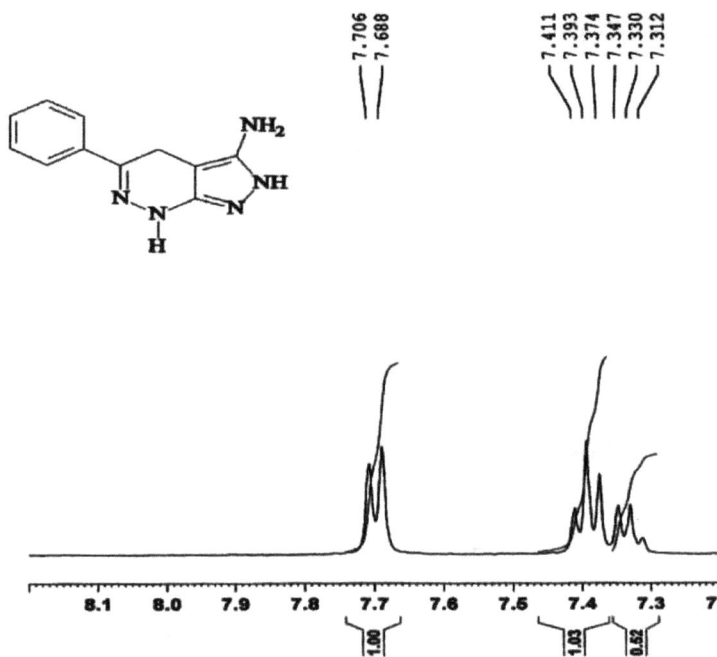

Fig. 118

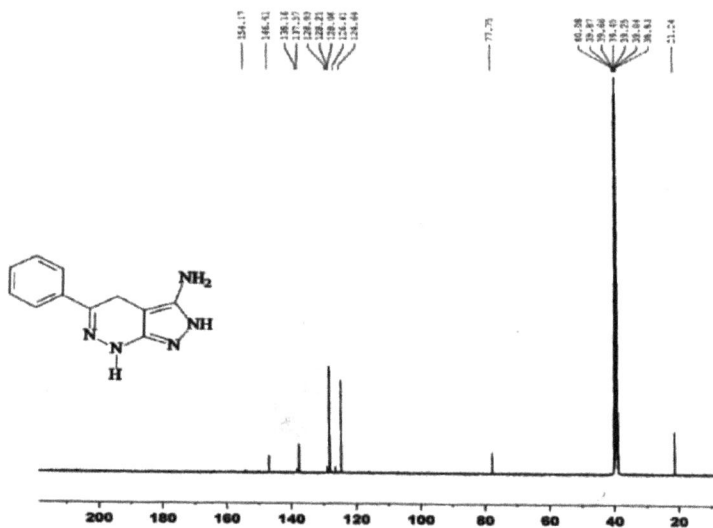

Fig. 119

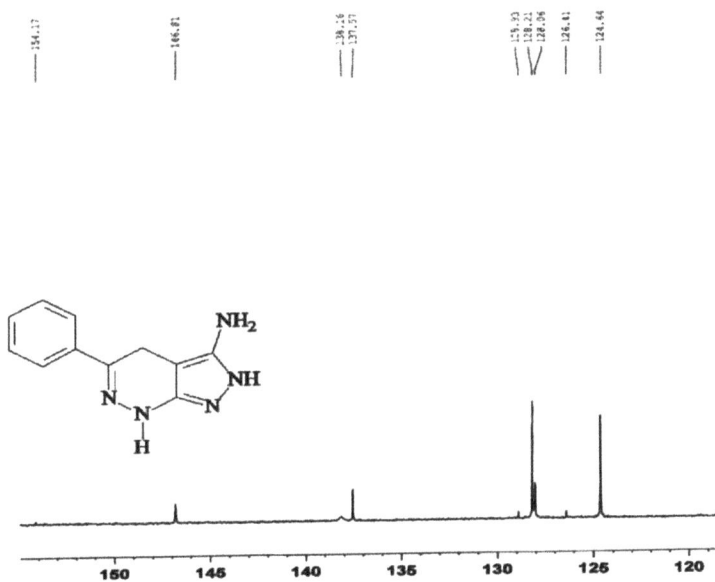

Fig. 120

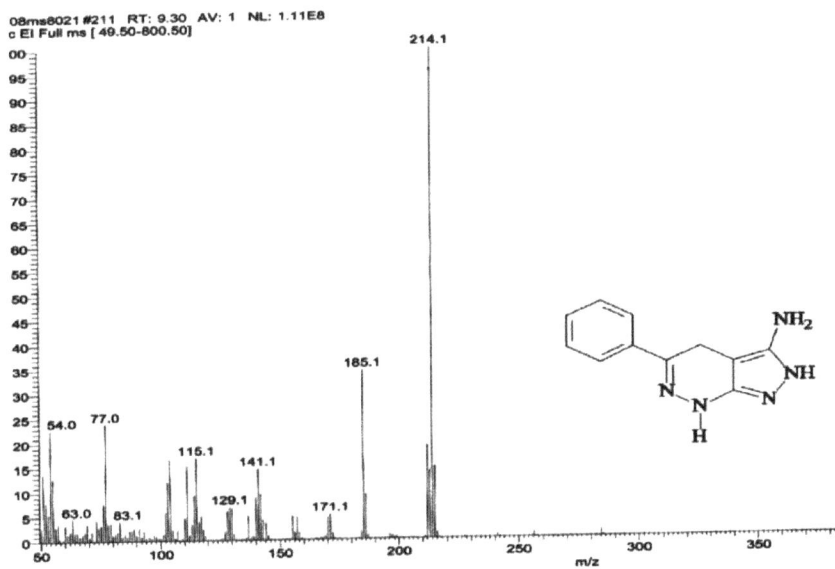

Fig. 121

Problem 12: The following reactions were repeatedly reported in the literature.

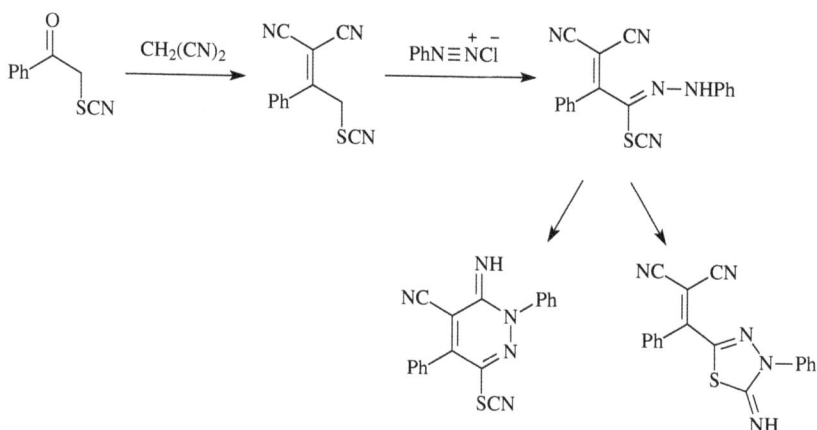

The stability of the formed arylhydrazone led us to reinspect these conclusions. In Figs. 122–159 are shown results that indicated the following final proof was got using X-ray.

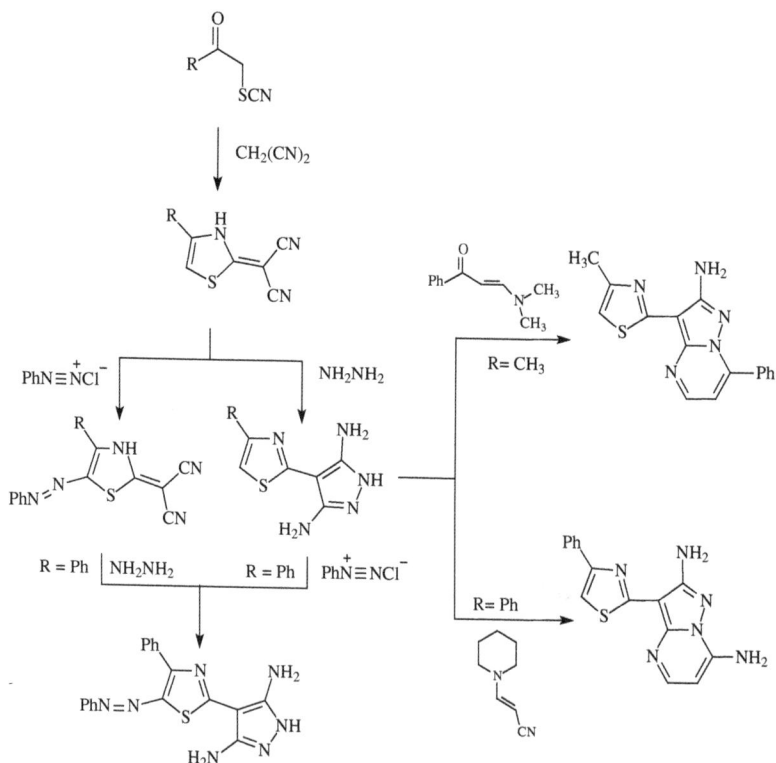

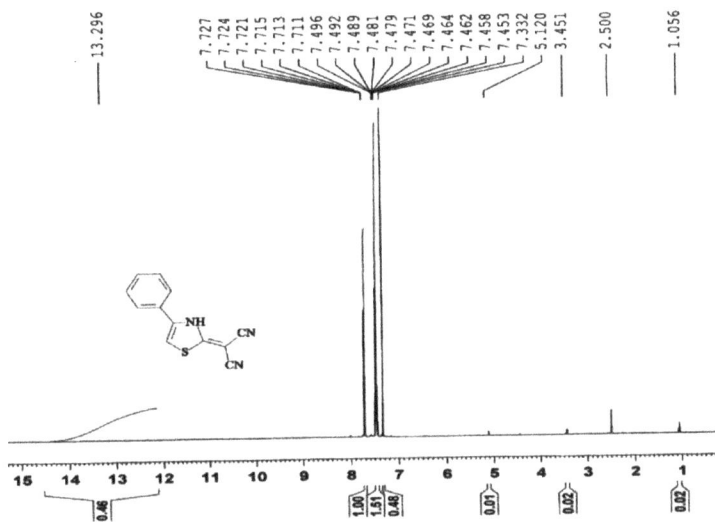

Fig. 122

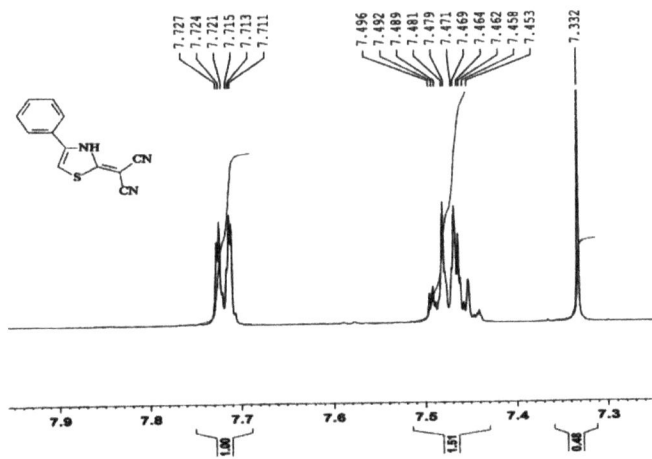

Fig. 123

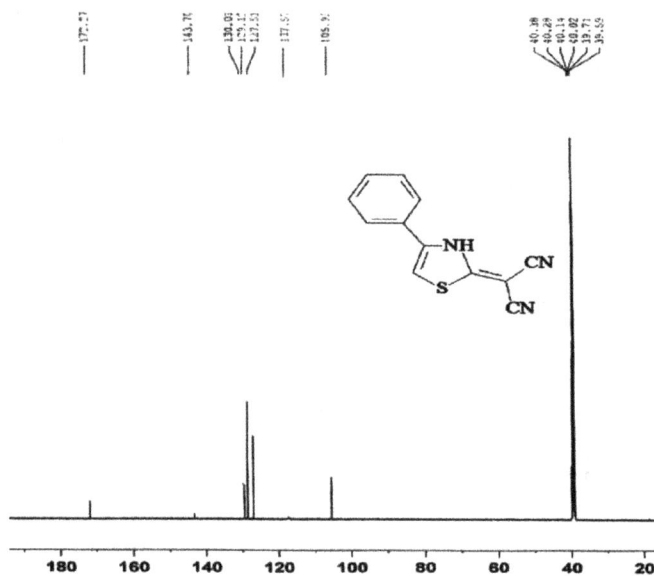

Fig. 124

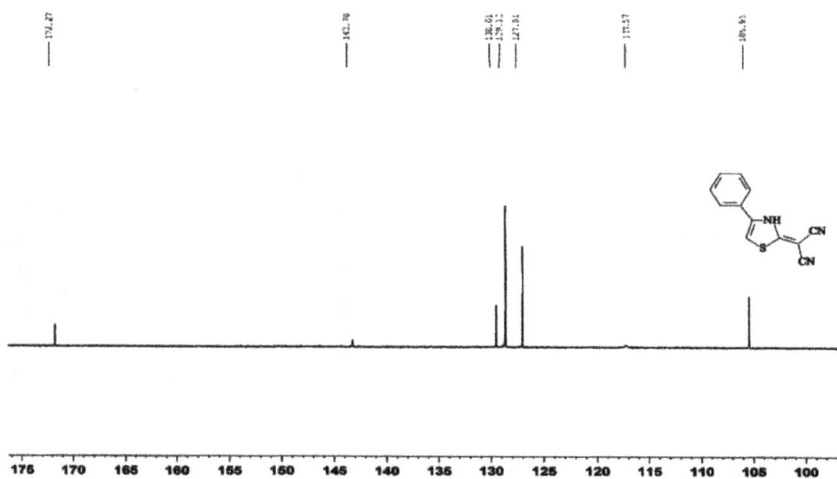

Fig. 125

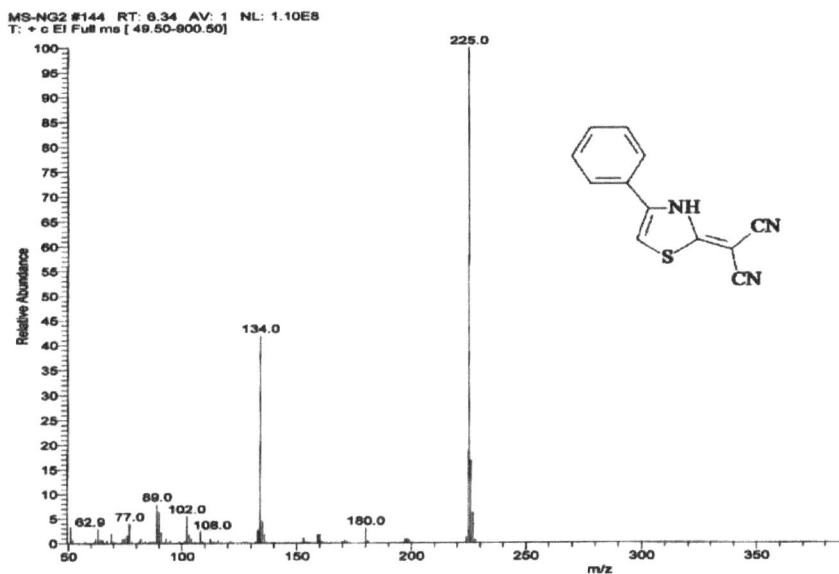

MS-NG2 #144 RT: 6.34 AV: 1 NL: 1.10E8
T: + c EI Full ms [49.50-900.50]

Fig. 126

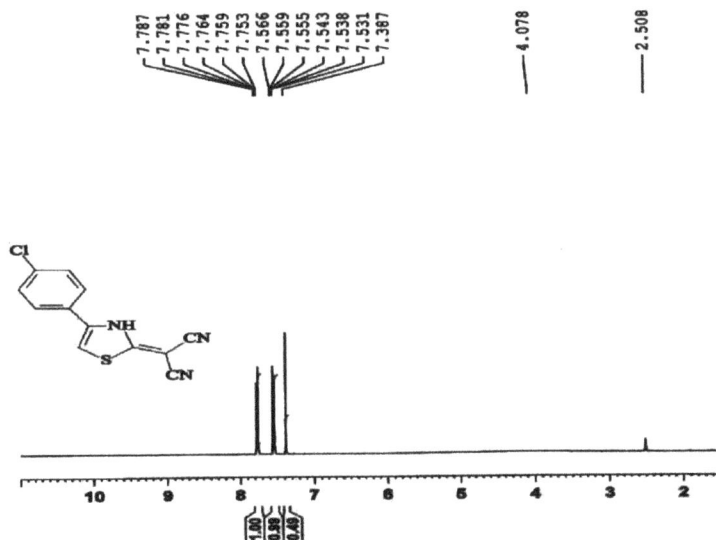

Fig. 127

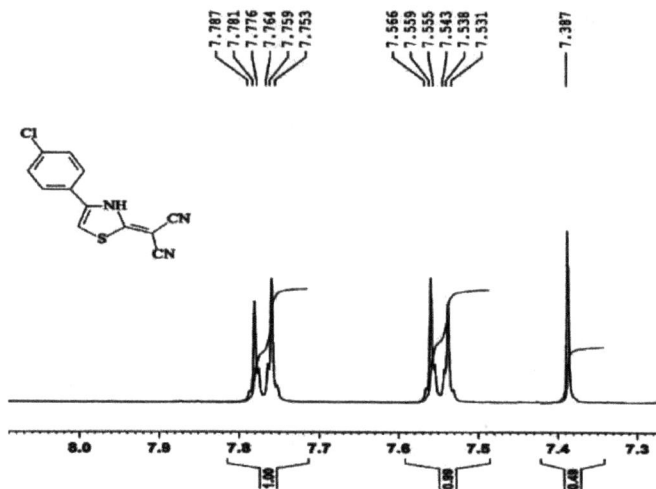

Fig. 128

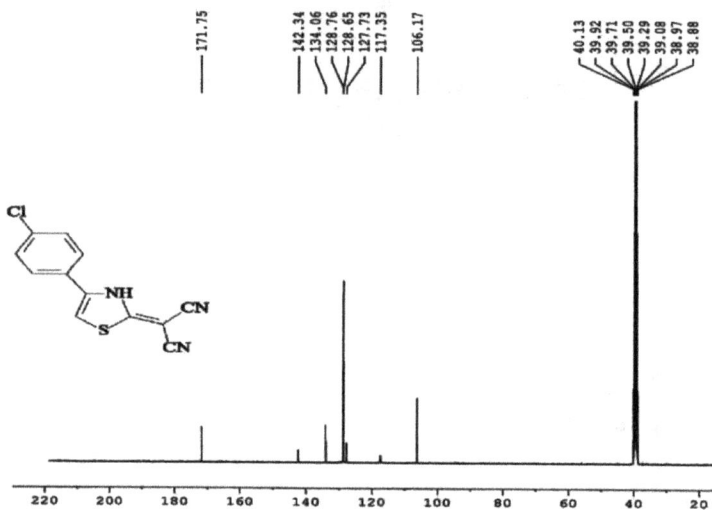

Fig. 129

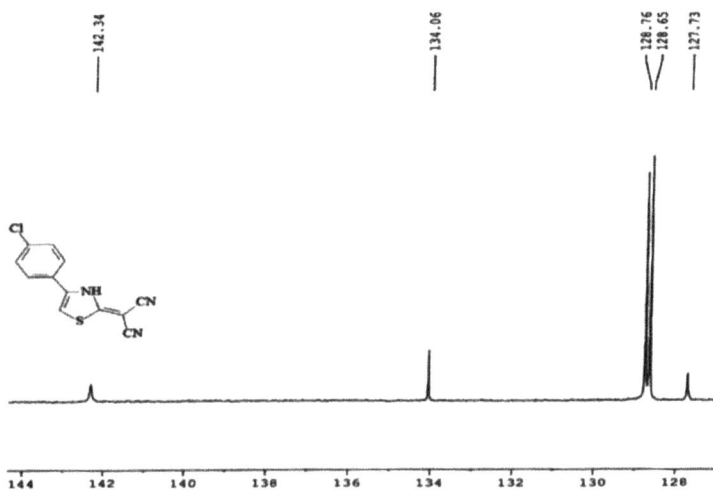

Fig. 130

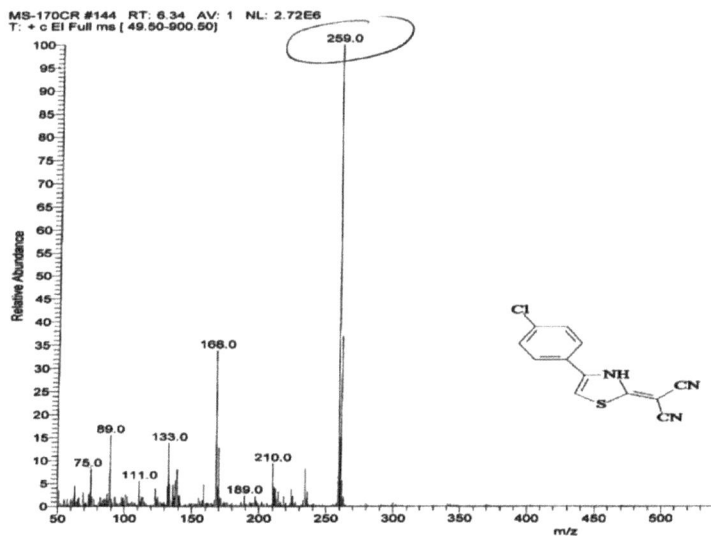

Fig. 131

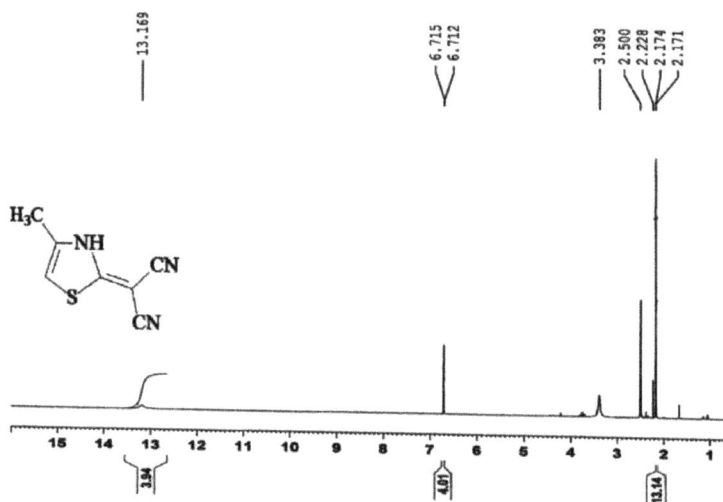

Fig. 132

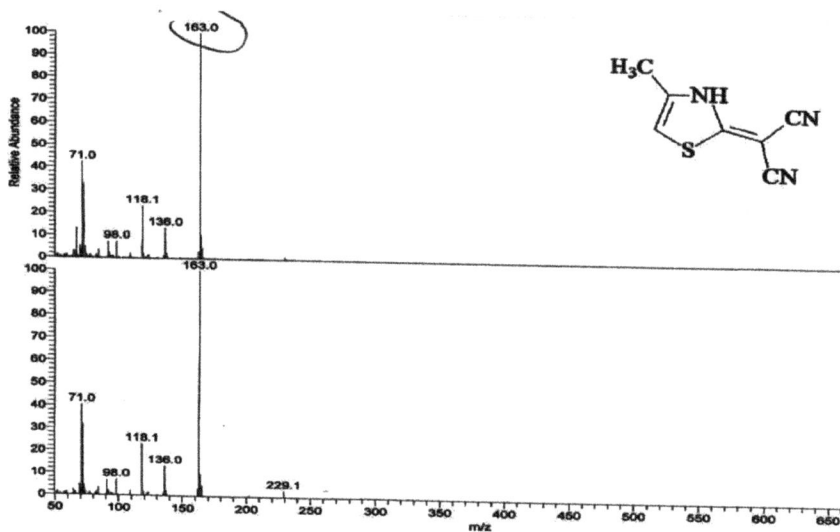

Fig. 133

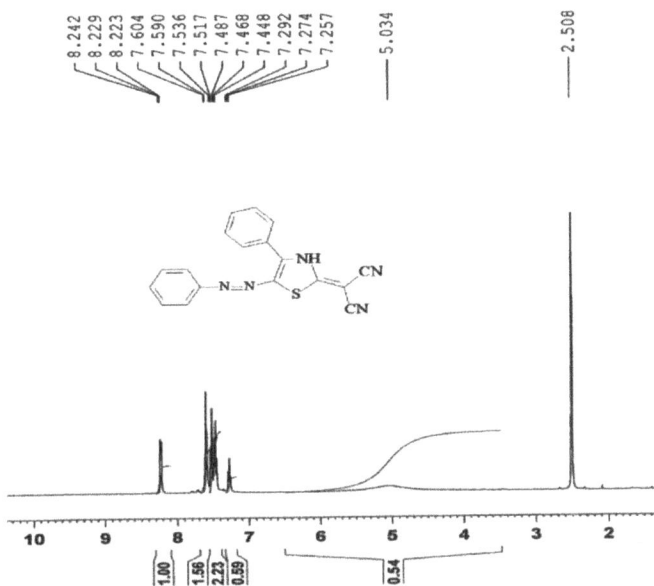

Fig. 134

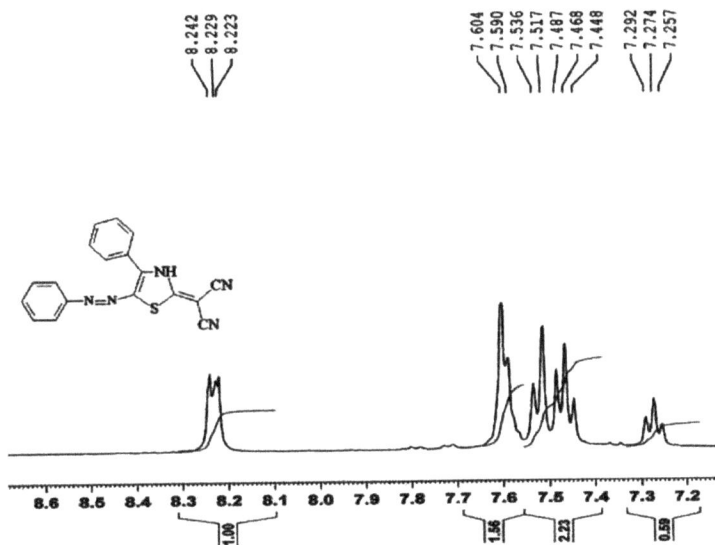

Fig. 135

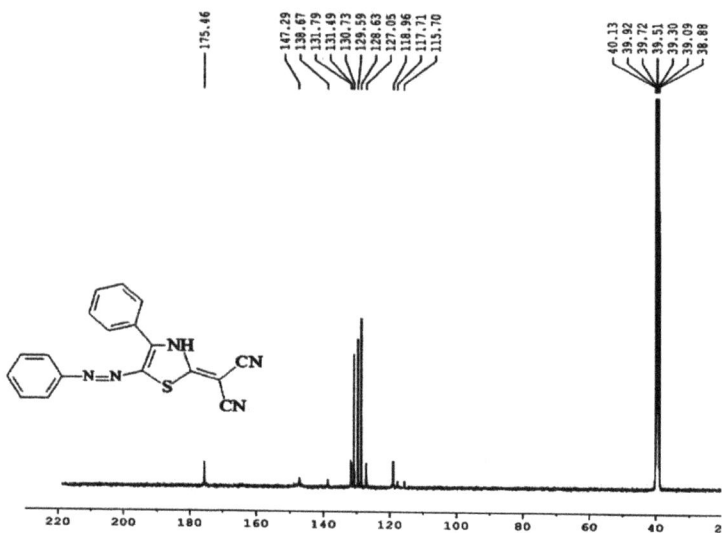

Fig. 136

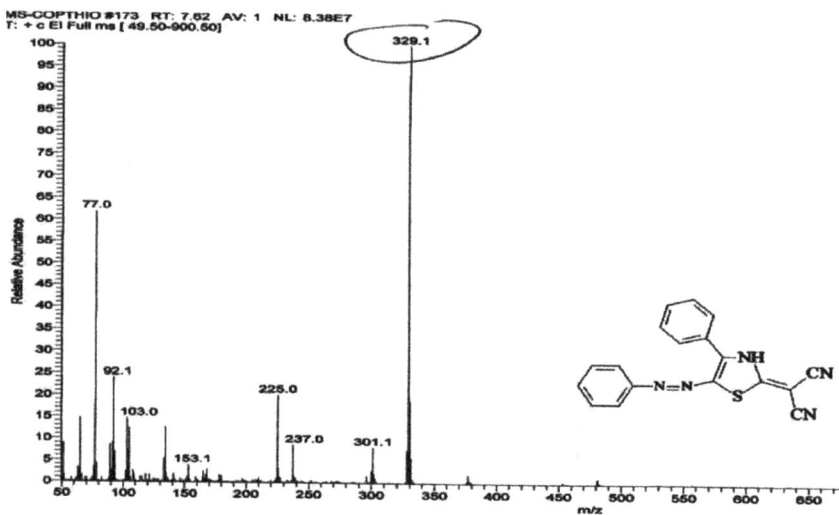

Fig. 137

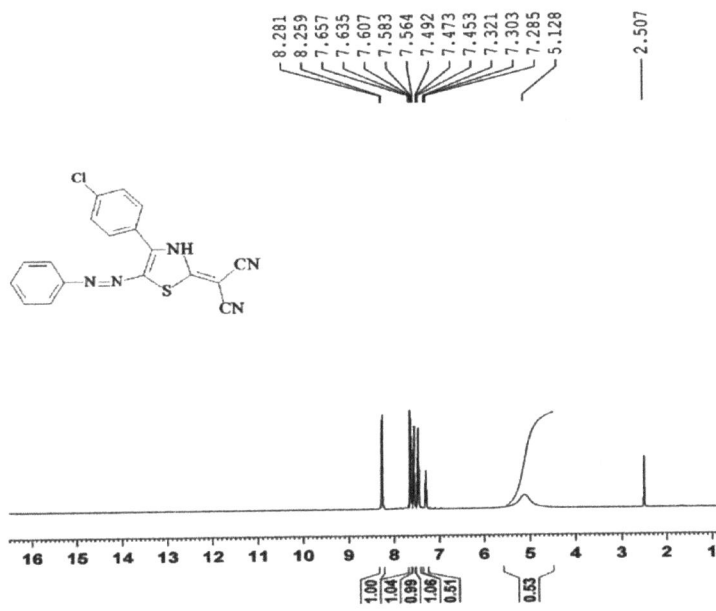

Fig. 138

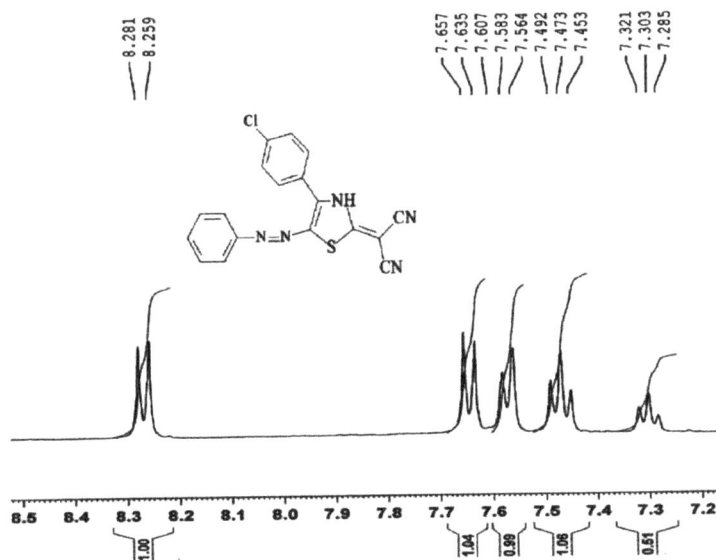

Fig. 139

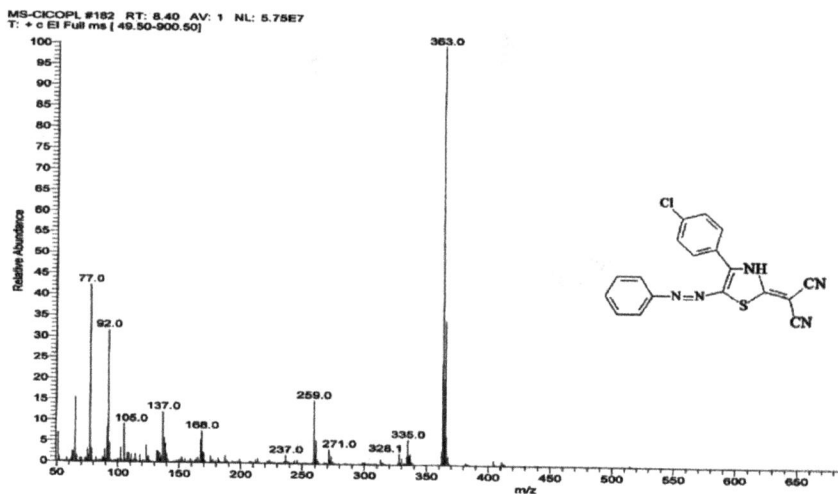

Fig. 140

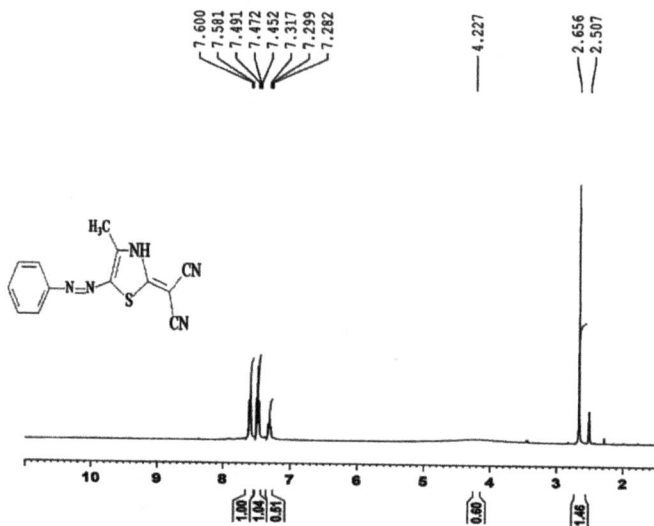

Fig. 141

7.600
7.581
7.491
7.472
7.452
7.317
7.299
7.282

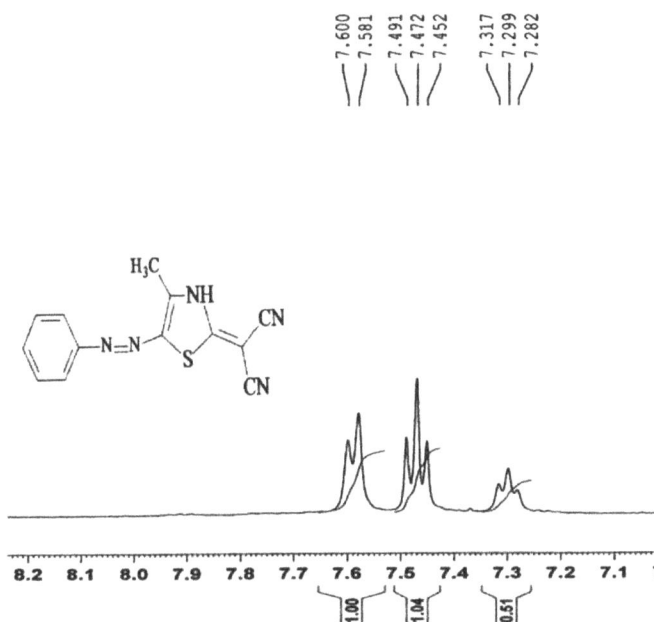

Fig. 142

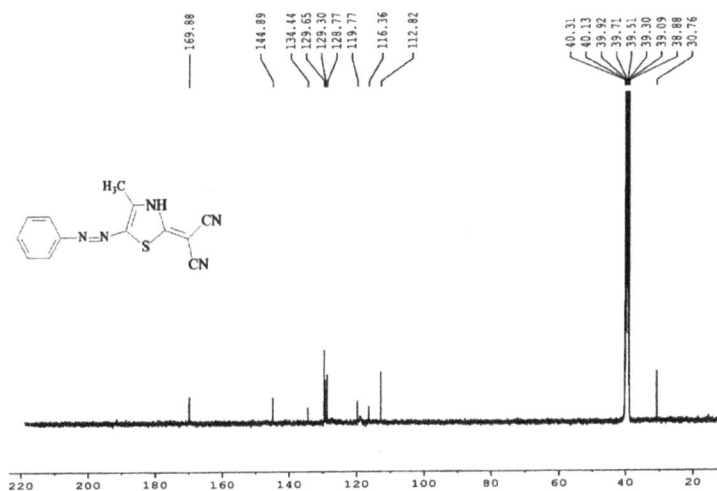

169.88

144.89
134.44
129.65
129.30
128.77
119.77
116.36
112.82

40.31
40.13
39.92
39.71
39.51
39.30
39.09
38.88
30.76

220   200   180   160   140   120   100   80   60   40   20

Fig. 143

MS-CICOPL #182 RT: 8.40 AV: 1 NL: 5.75E7
T: + c EI Full ms [49.50-900.50]

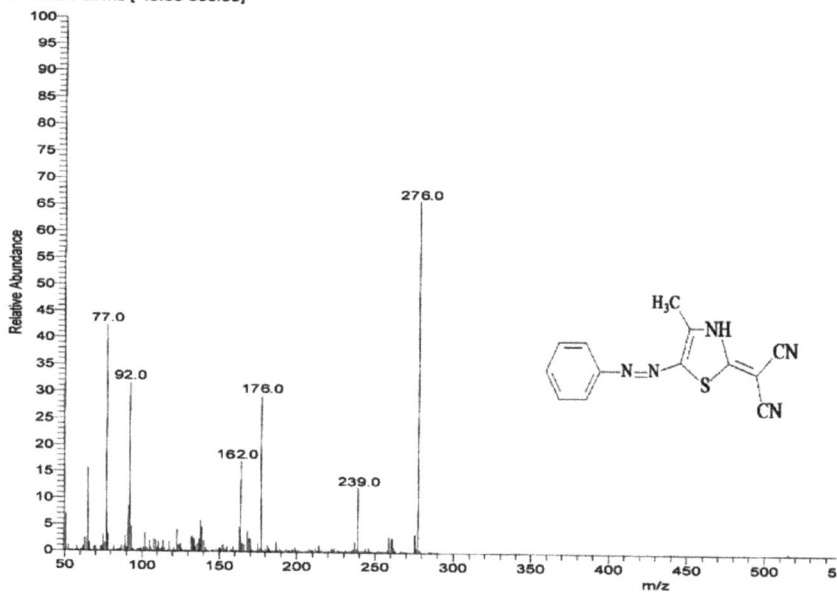

Fig. 144

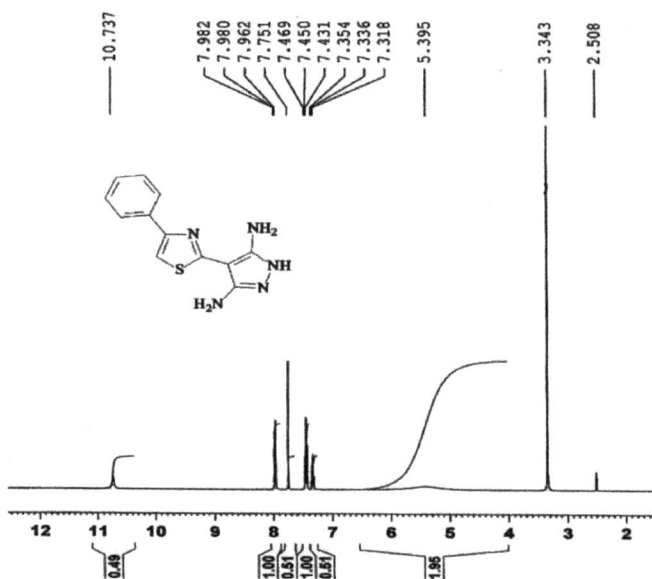

Fig. 145

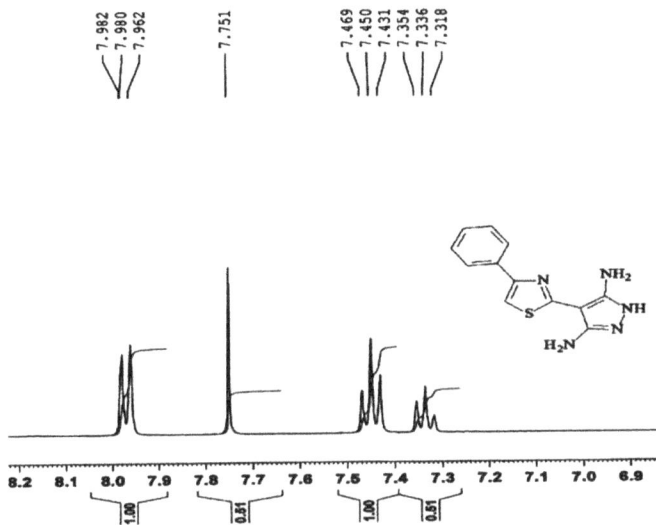

Fig. 146

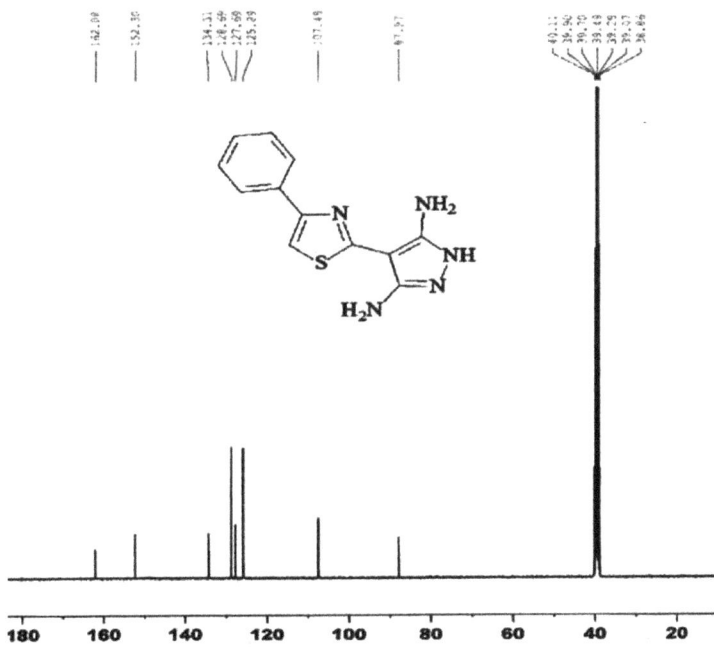

Fig. 147

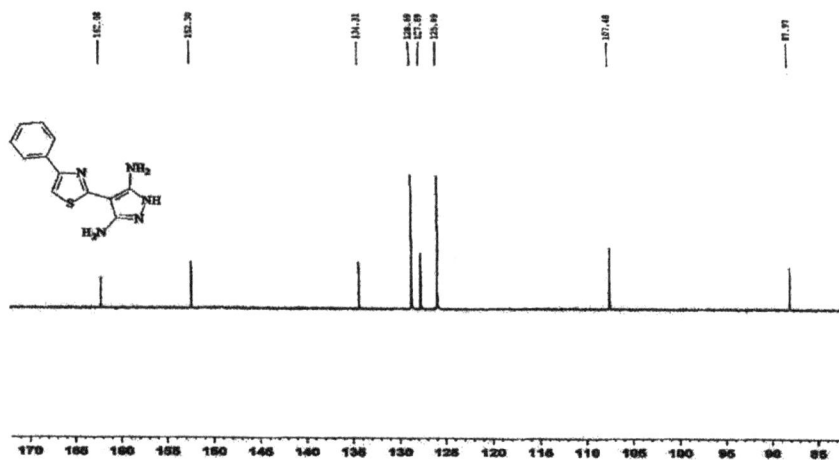

Fig. 148

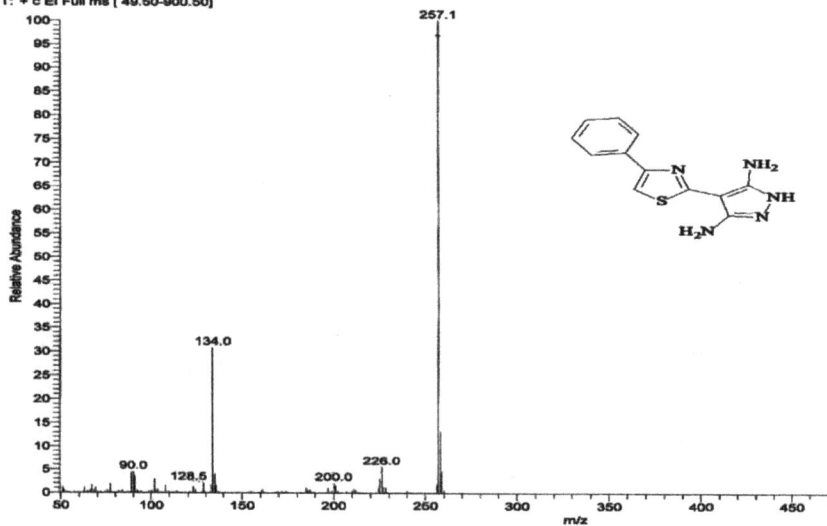

Fig. 149

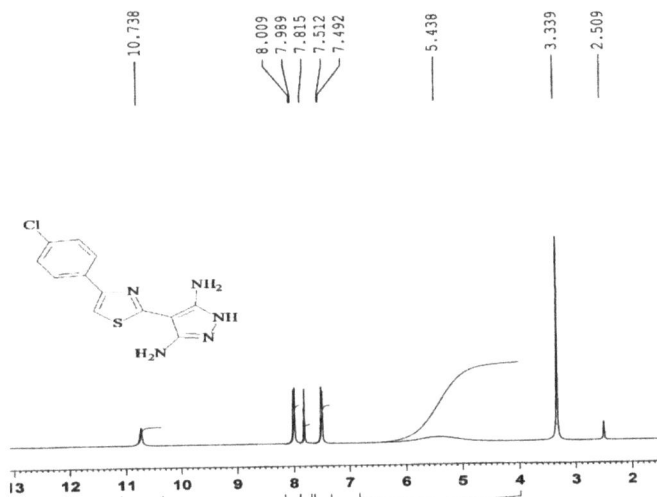

Fig. 150

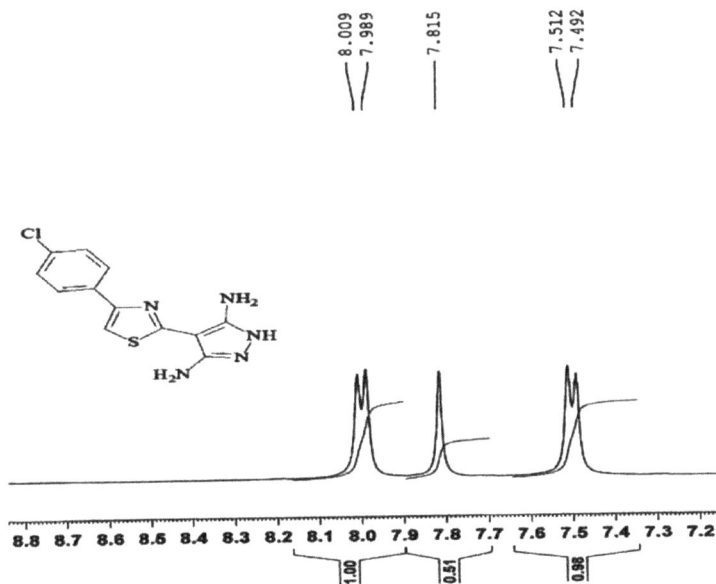

Fig. 151

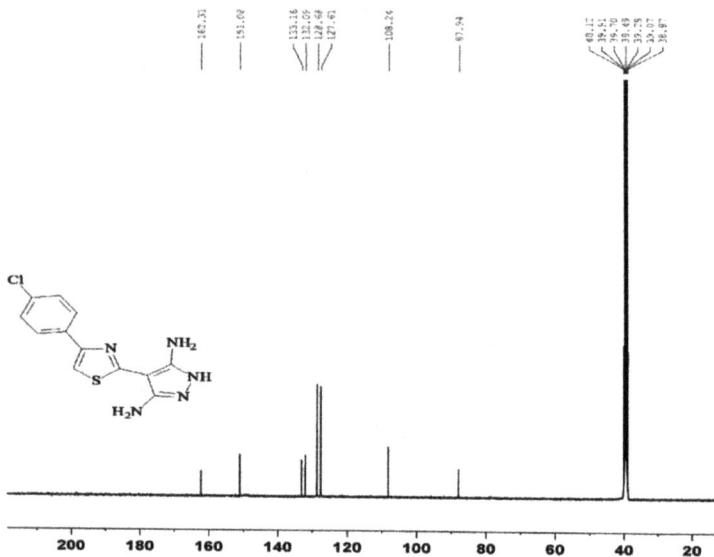

Fig. 152

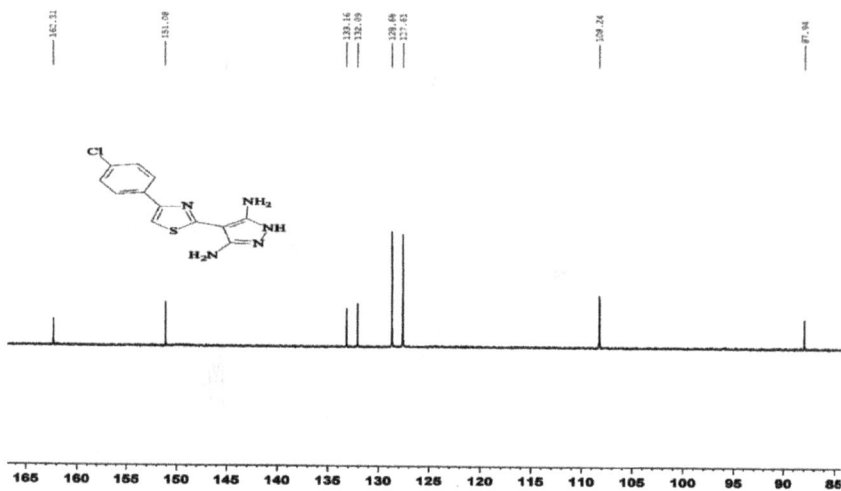

Fig. 153

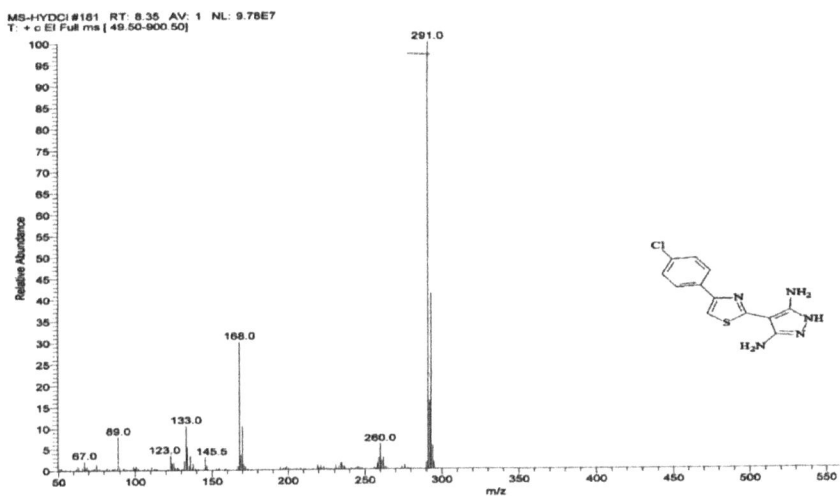

Fig. 154

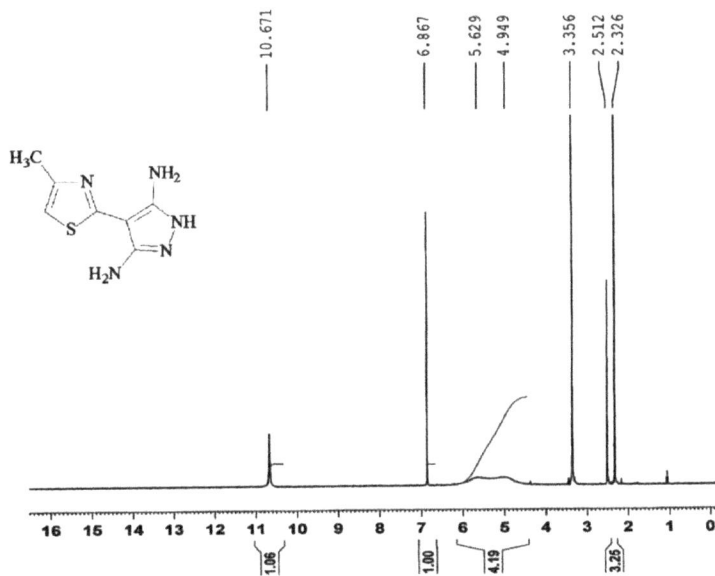

Fig. 155

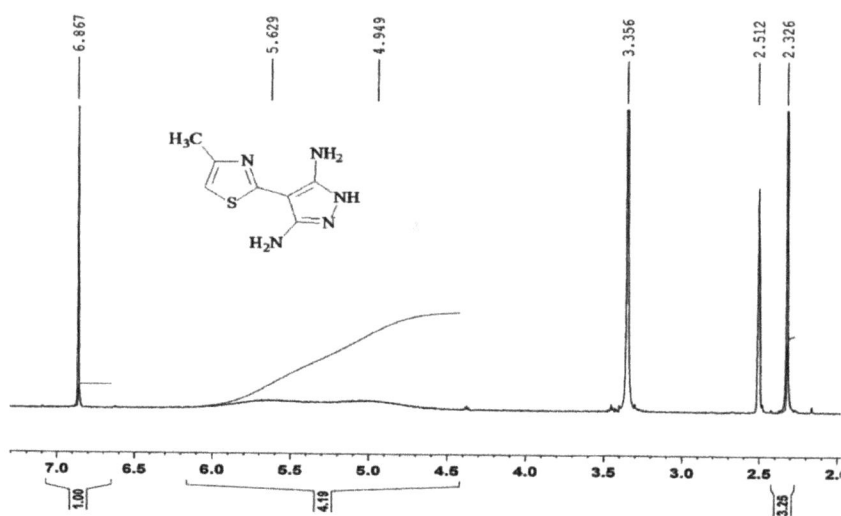

Fig. 156

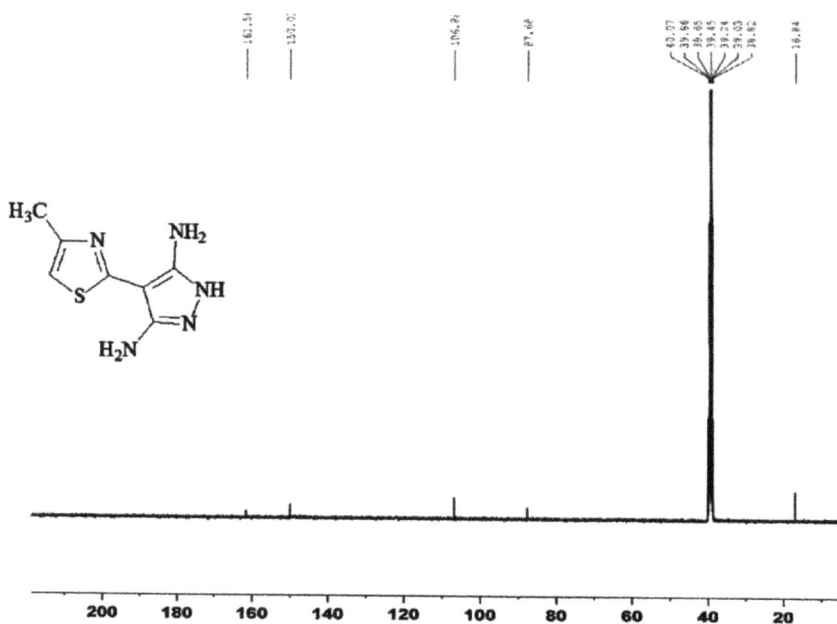

Fig. 157

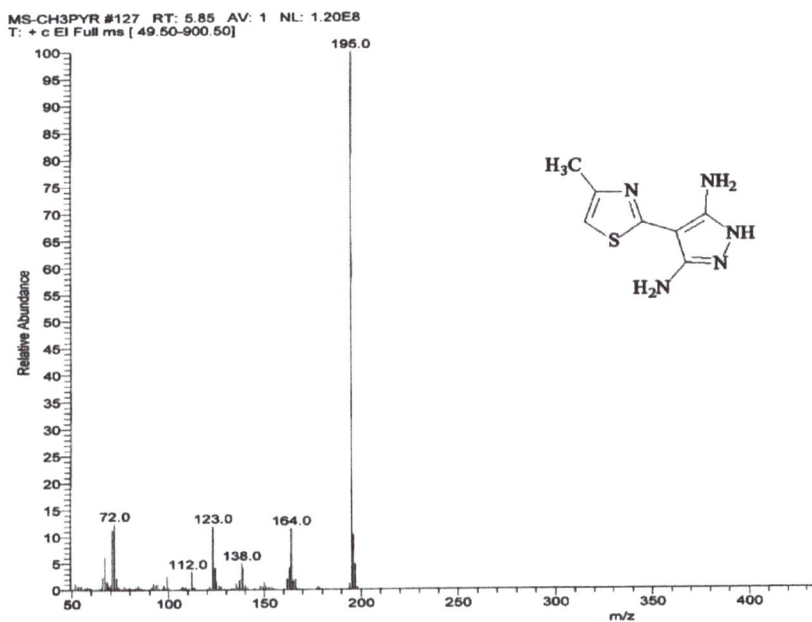

Fig. 158

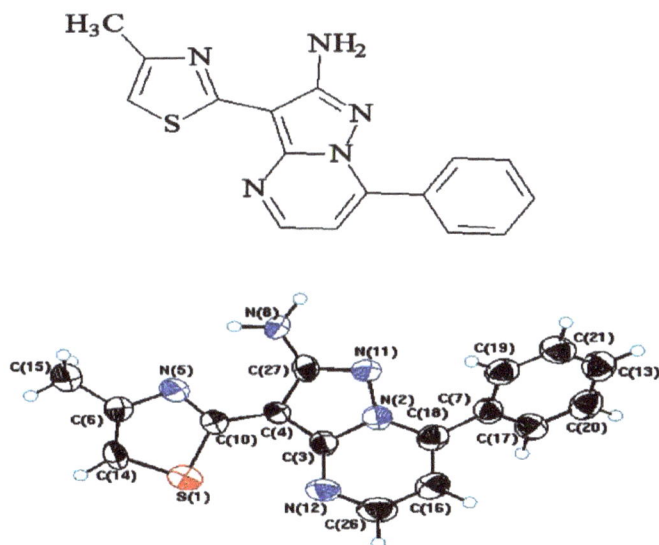

Fig. 159

References

1. M. H. Elnagdi, N. M. Abed, M. R. H. Elmoghayer and D. H. Fleta, *Indian J. Chem.*, *14B*, 422 (**1976**).
2. M. Quinterio, C. Seoane, J. L. Sotto, *Tetrahedron Lett.*, 1835 (**1977**).
3. J. Hollas, *Modern Spectroscopy*, John Wiley & Sons (**2004**).
4. C. N. Banwell and E. M. McCash, *Fundamental of Molecular Spectroscopy*, 5th Edn., McGraw-Hill Publisher (**2017**).
5. J. Keeler, *Understanding NMR Spectroscopy*, John Wiley & Sons (**2010**).
6. M. H. Elnagdi, A. M. Negm and A.W. Erian, *Liebigs Ann. Chem.* 1255 (**1989**).
7. M. H. Elnagdi, A. M. Negm, and K. U. Sadek, *Synlett.* 27 (**1994**).
8. M. H. Elnagdi and A.W. Erian, *Liebigs Ann. Chem.* 1215 (**1990**).
9. M. M. Abdelkhalik, A. M. Negm, A. I. Elkhouly and M. H. Elnagdi, *Heteroatom. Chem.* 502 (**2004**).
10. E. Nyiondi-Bonguen, E. S. Fondigo, Z. T. Famam and D. Doepp, *J. Chem. Soc. Perkin Trans.* 1, 2191 (**1994**).
11. F. All-Omran, M. M. Abdelkhalek, H. Al-Awadi and M. H. Elnagdi, *Tetrahedron*, *52*, 11915 (**1996**).
12. A. Al-Etiabi, N. Al-Awadi, F. Al-Omran, M. M. Abdel-Khalik and M. H. Elnagdi, *J. Chem. Res. (S)*, 151 (**1990**).
13. E. S. Fondjo, D. Döpp and G. Henkel, *Tetrahedron*, *62*, 7121 (**2006**).
14. G. E. H. Elgemeie, M. M. Sallam, S. M. Sherif and M. H. Elnagdi, *Heterocycles*, *23*, 3107 (**1985**).
15. F. M. Abdelrazek, Z. E. Kandeel, K. M. H. Hilmy and M. H. Elnagdi, *Synthesis* 432 (**1985**).
16. N. S. Ibrahim, Ph.D. Thesis, Cairo University (**1980**).
17. S. M. Al-Mousawi, K. Kaul, M. A. Mohamed and M. H. Elnagdi, *J. Chem. Res.* 318 (**1997**).
18. M. A. Al-Shiekh, A. M. S. El-Din, E. A. Hafez and M. H. Elnagdi, *J. Chem. Res.* 174 (**2004**).
19. H. F. Anwar and M. H. Elnagdi, *Arkivoc*, *i*, 198 (**2009**).
20. H. F. Anwar, D. H. Fleita, H. Kolshorn, H. Meier and M. H. Elnagdi, *Arkivoc*, *xi*, 133 (**2006**).
21. M. A. Al-Shiekh, H. Y. Medrassi, M. H. Elnagdi and E. A. Hafez, *Arkivoc*, *xvii*, 36 (**2008**).

4 Diastereotopic Protons

4.1. An Asymmetric Synthesis Proposal for Synthesis of Biologically Relevant Heterocycles

In plenty of our previous works, we created a chiral center. As examples, see Schemes 1–3. Usually, the products we got were racemates as the reactions did not involve a chiral reagent. How is one to perform asymmetric synthesis in light of the interest in this type of synthesis. The simplest way is by using a chiral catalyst. Then comes the next question: how to precisely calculate diastereomeric or enantiomeric excess? Inspection of integrals of diastereotopic protons seems to be the ideal way. But, what are diastereotopic protons. These are methylene (CH_2) groups adjacent to or a little farther away from the created chiral centers. And if one of those methylenes was replaced by deuterium, another chiral center may be produced; thus, these methylenes are magnetically different. Unfortunately, until recent, many authors were unaware of this fact and reported these groups on ^1H NMR spectra as the simple CH_2. One can just read through the literature to see this. In the following image, these spectra are demonstrated.[1]

151

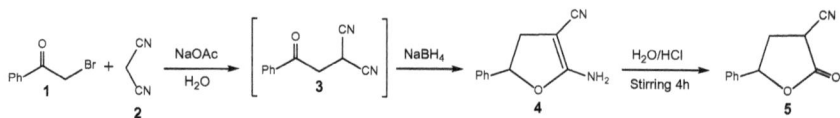

Scheme 1

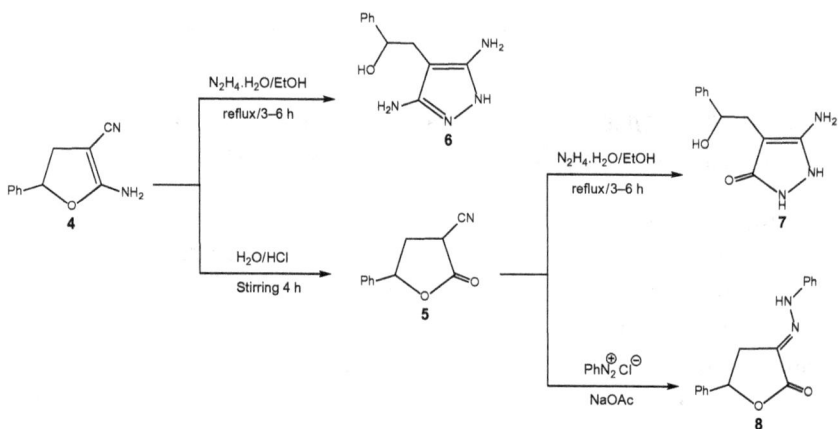

Scheme 2

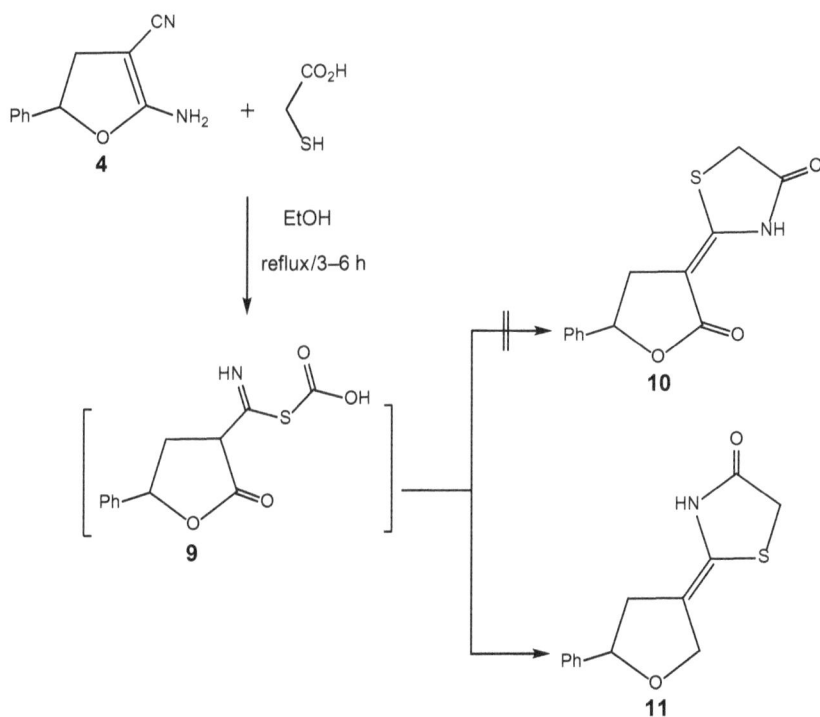

Scheme 3

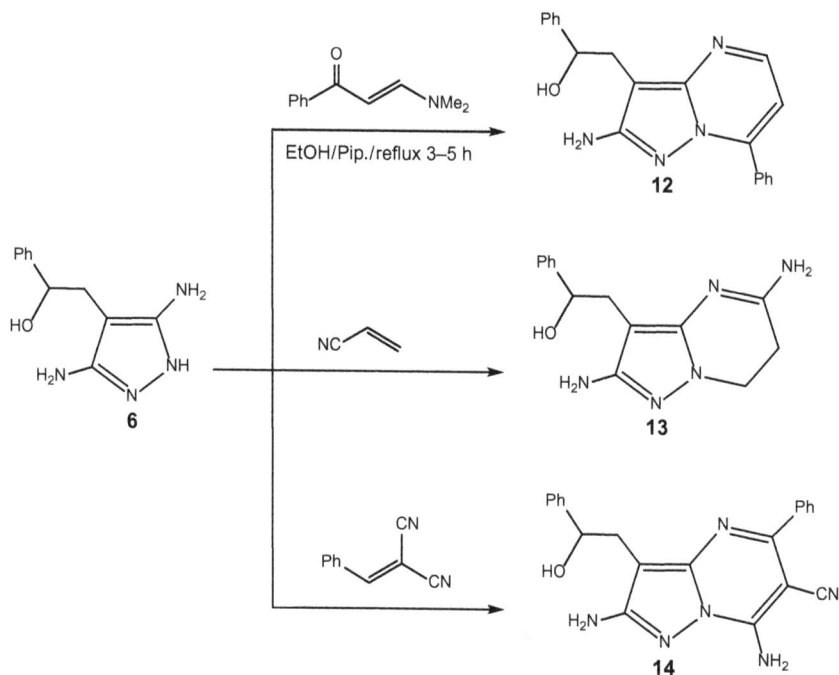

Scheme 4

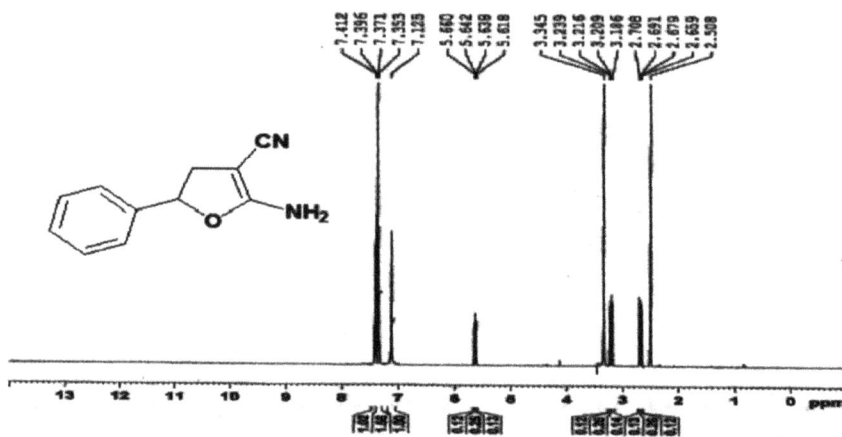

Fig. 1 ^{1}HNMR of compound **4** with diastereotopic protons at C-4

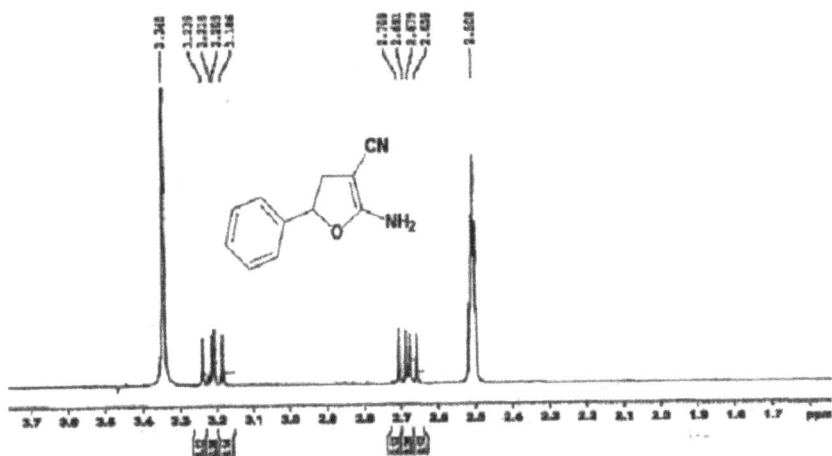

Fig. 2 Enlarged ^{1}HNMR of compound **4**

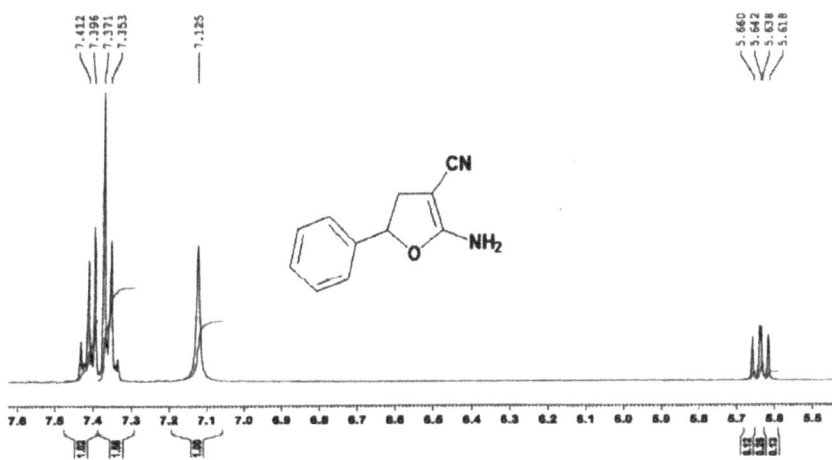

Fig. 3 Enlarged ^{1}HNMR of comounnd **4**

Spectroscopic Identification of Organic Molecules

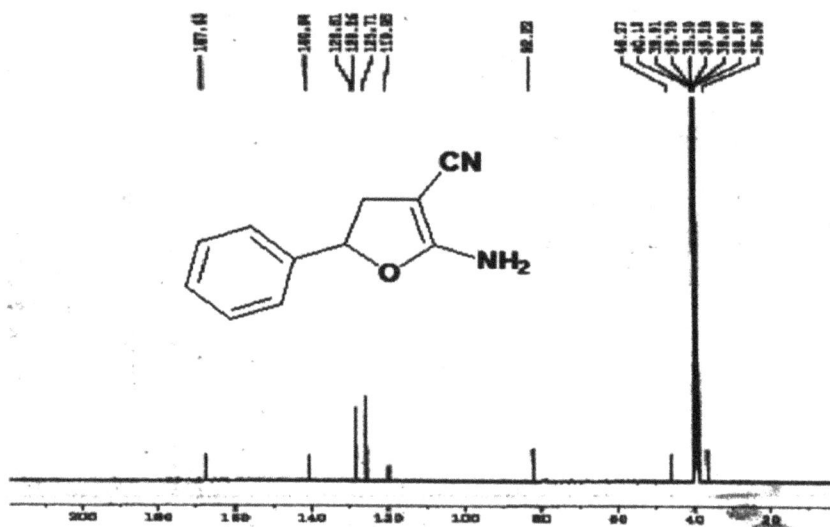

Fig. 4 ^{13}CNMR of compound **4**

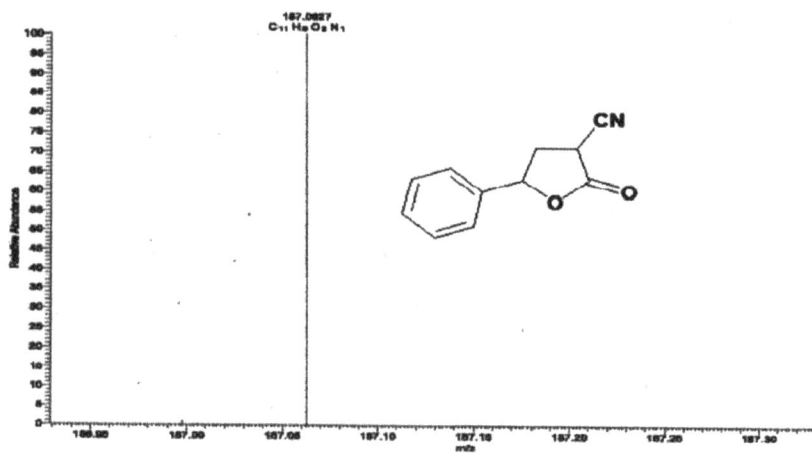

Fig. 5   Exact mass of compound **5**

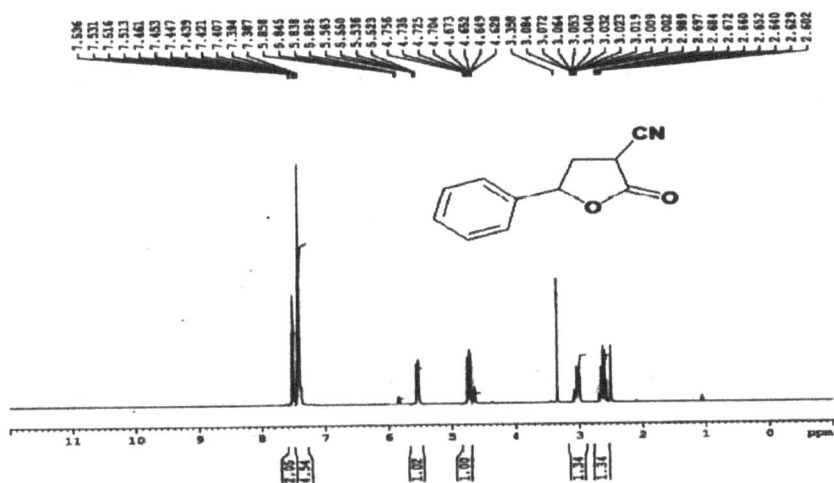

Fig. 6 ¹HNMR of compound **5** with diastereotopic protons at C-4

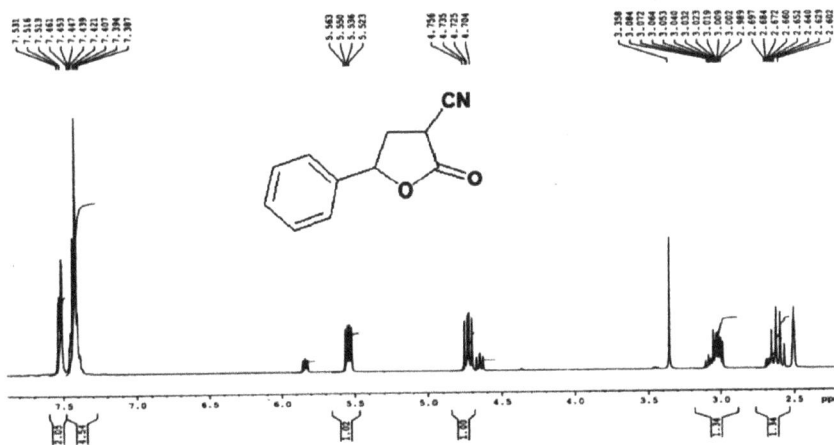

Fig. 7 Enlarged ¹HNMR of compound **5**

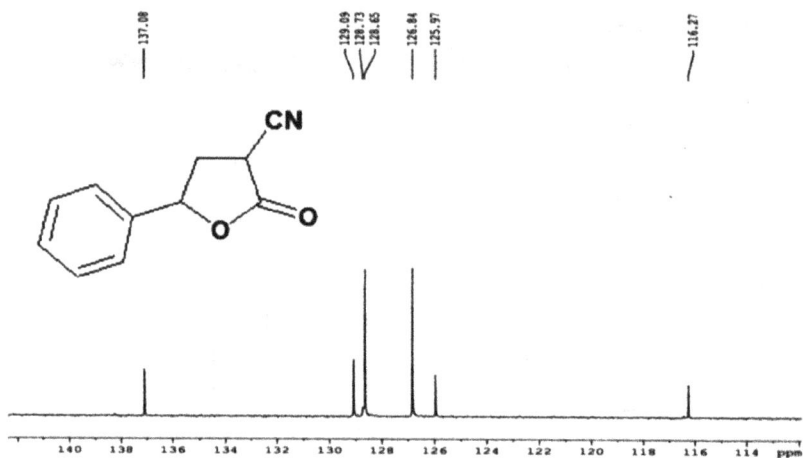

Fig. 8 ^{13}CNMR of compound **5**

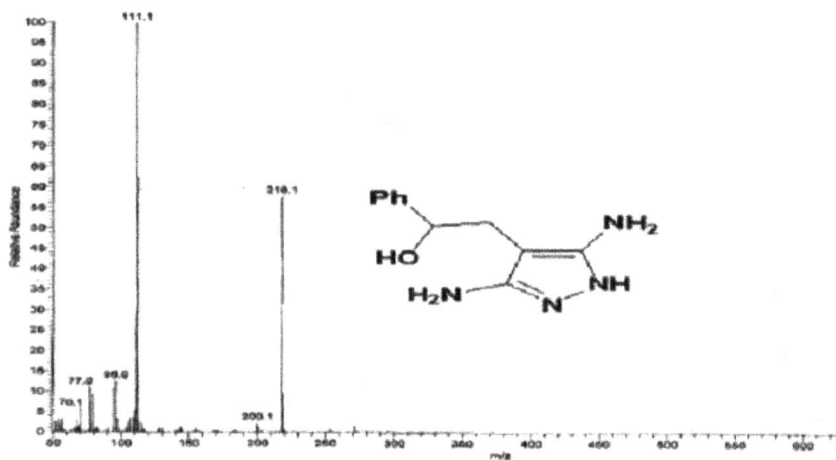

Fig. 9   Mass spectra of compound **6**

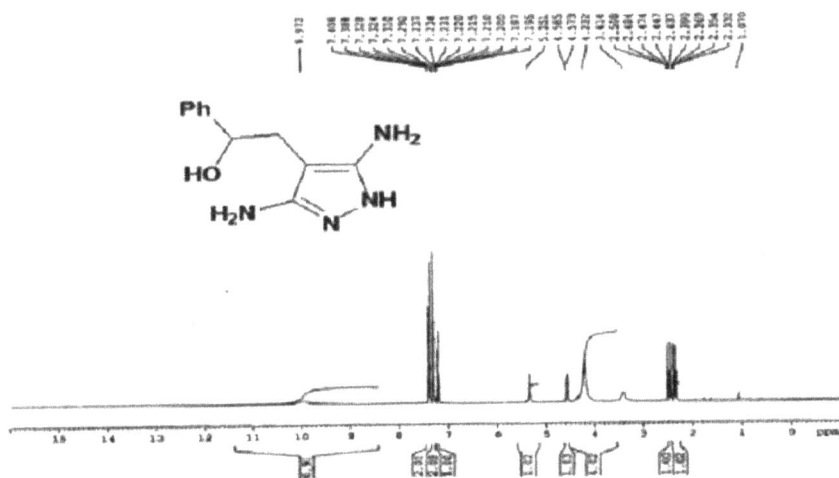

Fig. 10 ^{1}HNMR of compound **6** with the diastereotopic protons of . CH$_2$ group at C-4 phenylethyl group

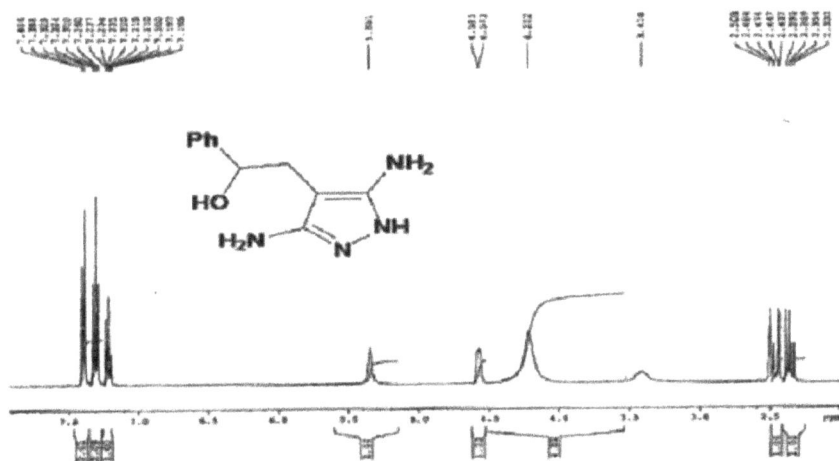

Fig. 11 Enlarged ^{1}HNMR of compound **6**

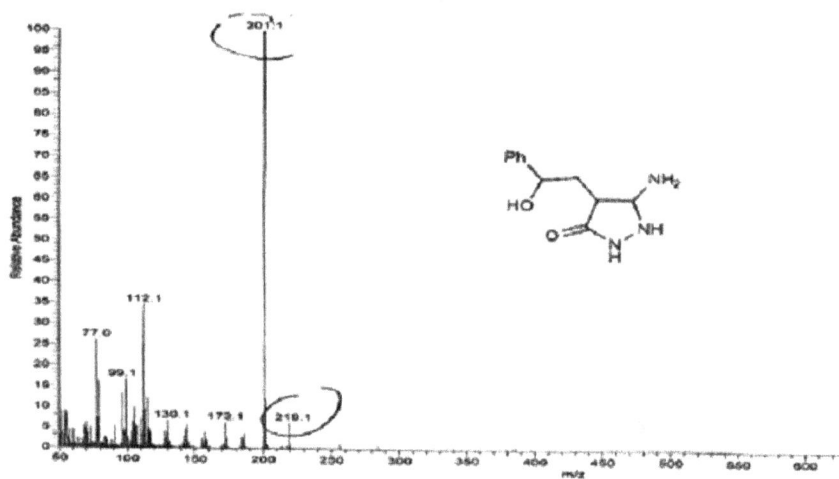

Fig. 12   Mass spectra of compound **7**

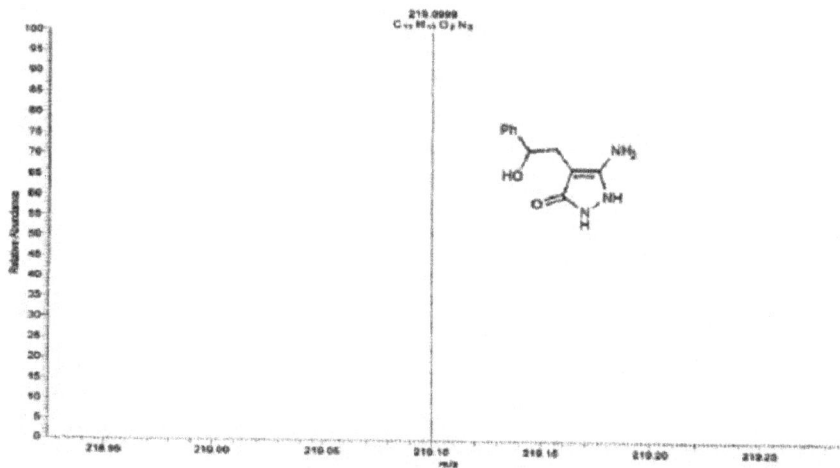

Fig. 13   Exact Mass of compound **7**

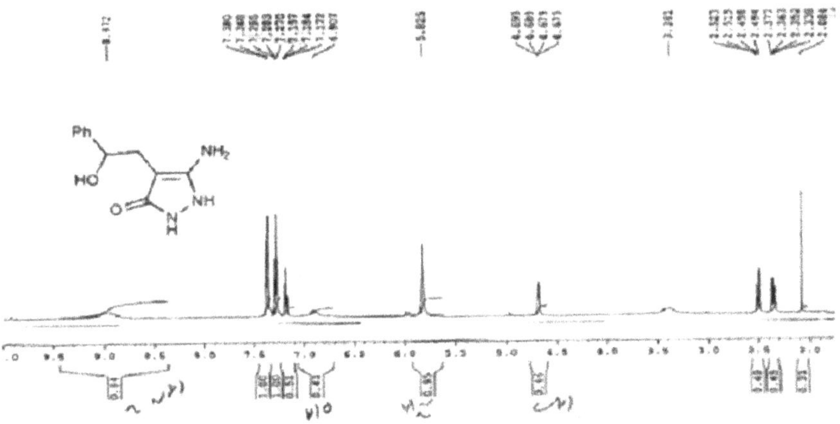

Fig. 14 ^{1}HNMR of compound **7** with diastereotopic protons of CH_2 at C-4 of phenylethyl group

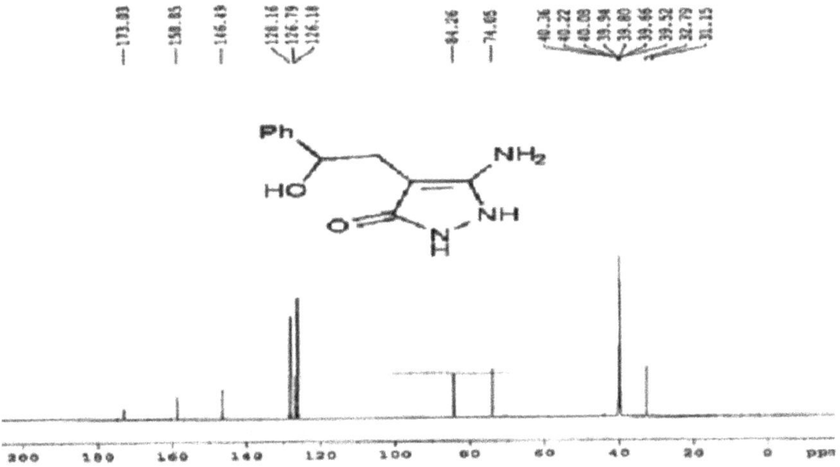

Fig. 15 ^{13}CNR of compound **7**

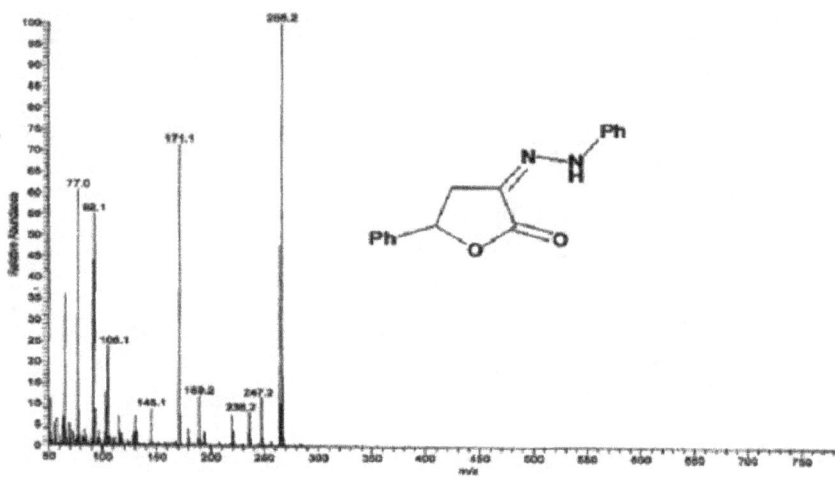

Fig. 16 Mass spectra of compound **8**

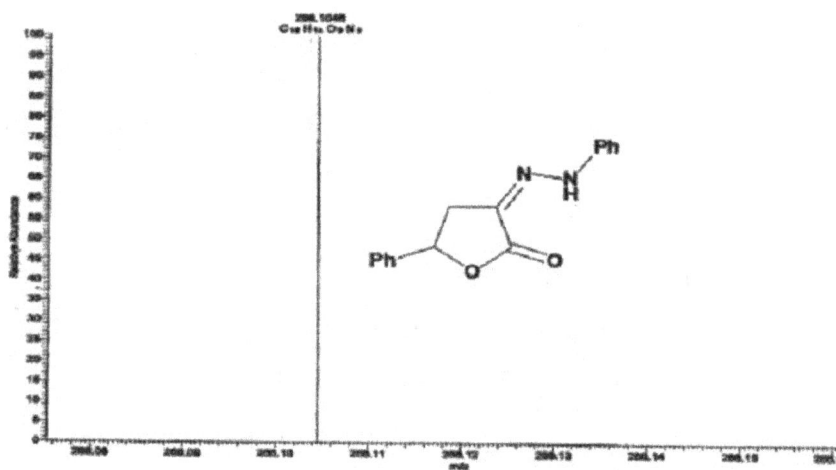

Fig. 17 Exact mass of compound **8**

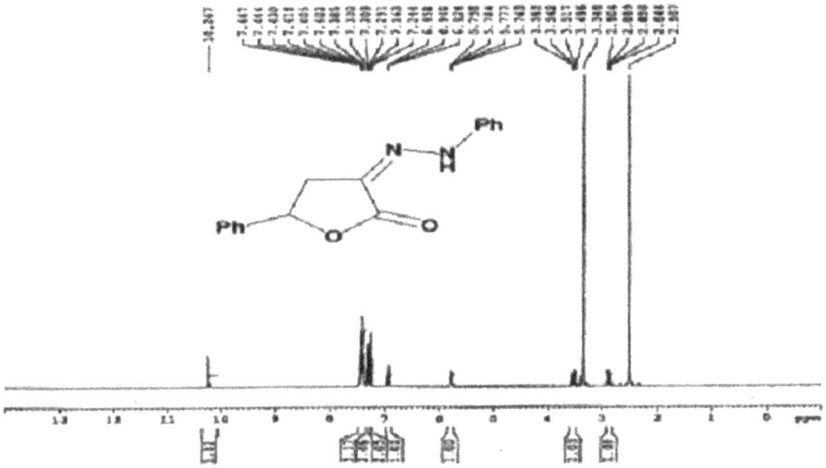

Fig. 18 ^{1}HNMR for compound **8** with diastereotopic protons of CH$_2$ of CH-4 phenylethyl group

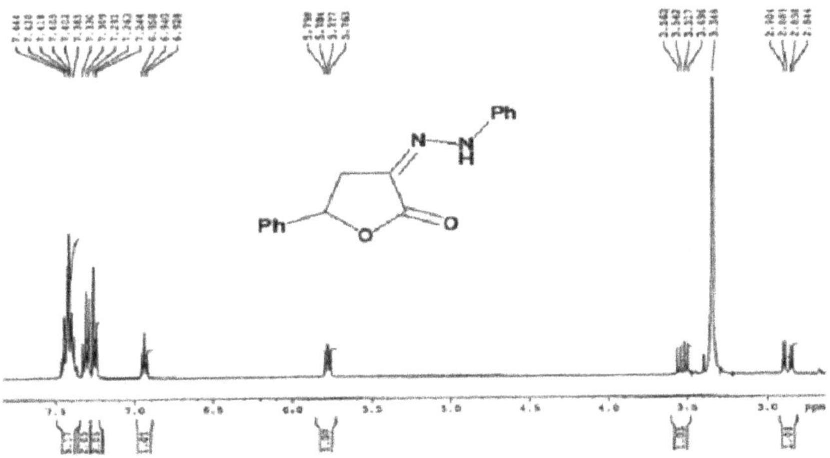

Fig. 19 Enlarged ^{1}HNMR for compound **8**

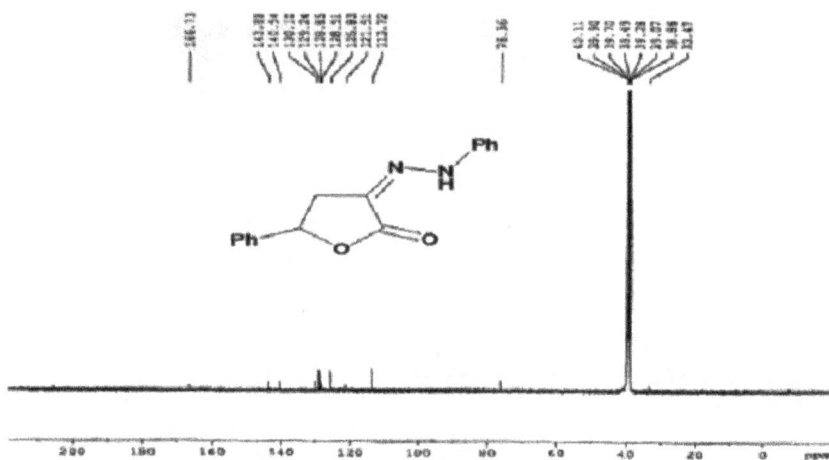

Fig. 20 ^{13}CNMR of compound **8**

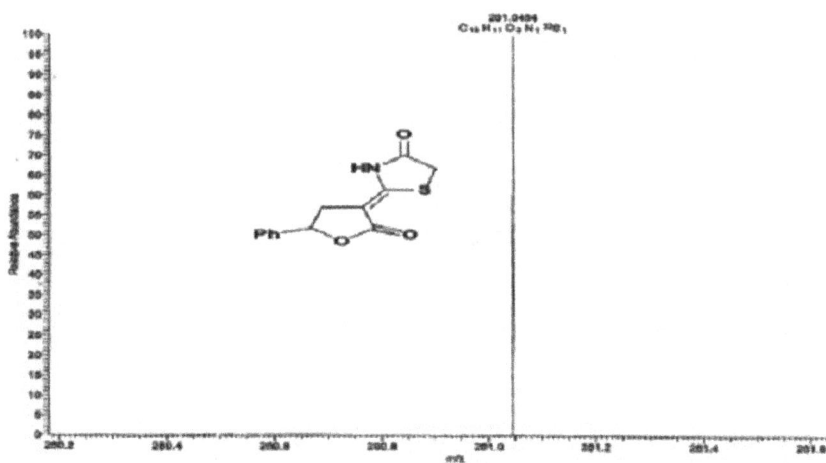

Fig. 21   Exact mass of compound **11**

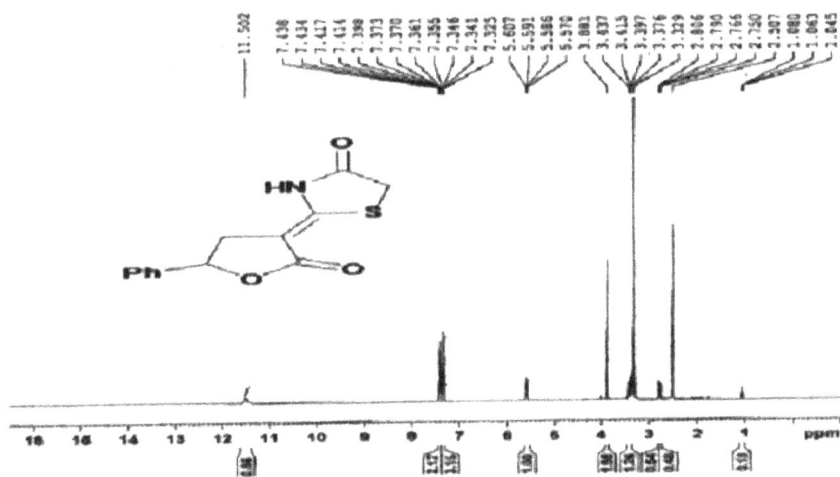

Fig. 22 ^{1}HNMR for compound **11** with diastereotopic protons of furan C-4

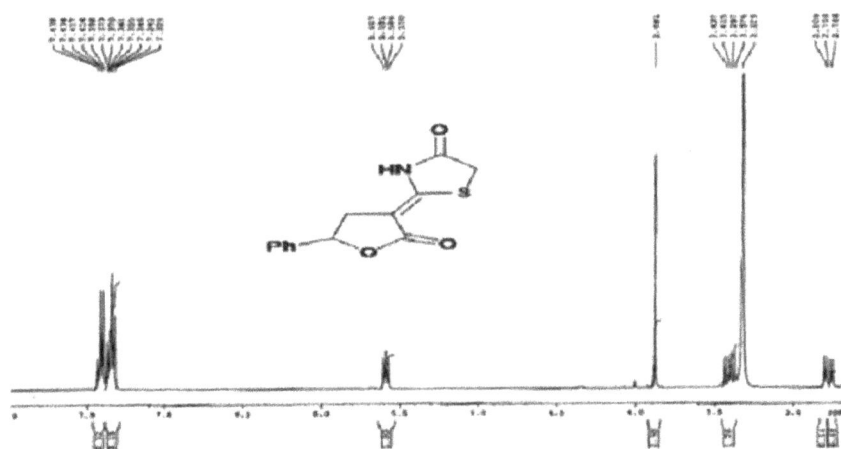

Fig. 23 Enlarged ^{1}HNM for compound **11**

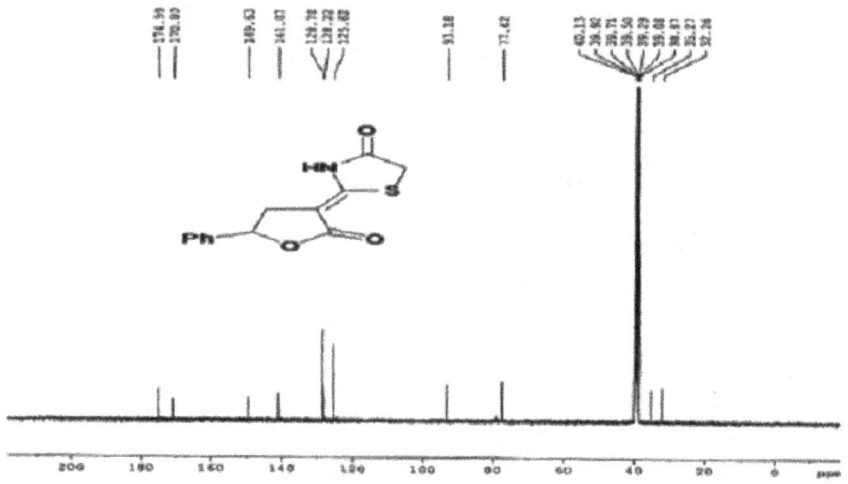

Fig. 24 ^{13}NMR for compound 11

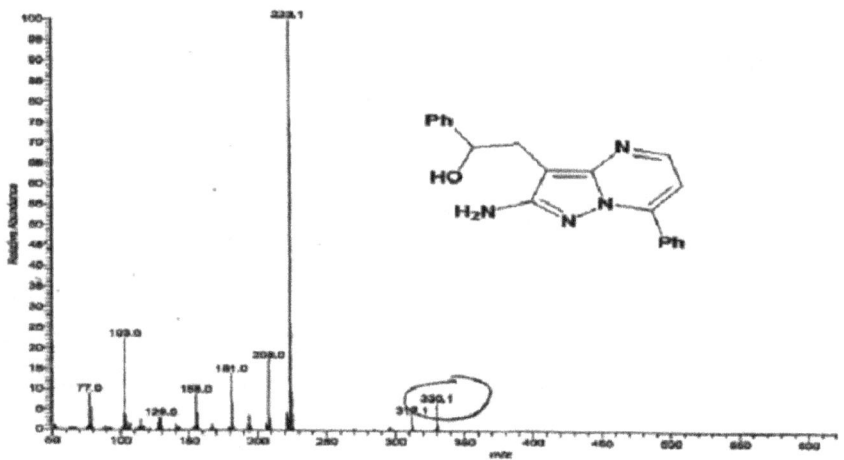

Fig. 25   Mass spectra of compound 12

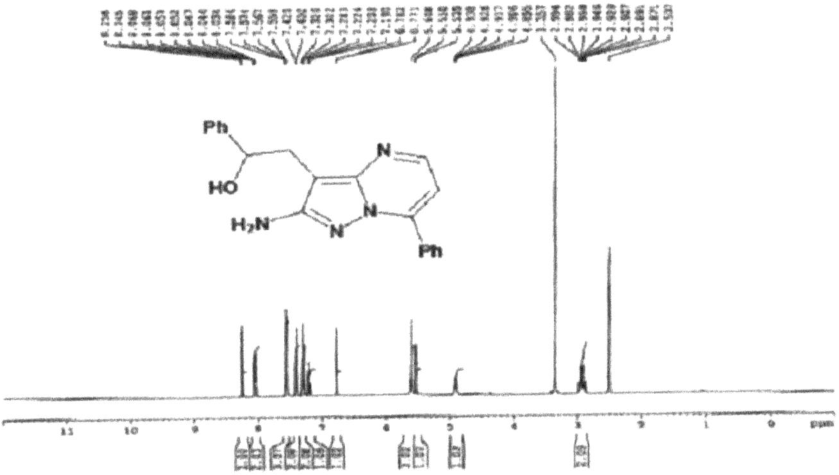

Fig. 26 ¹HNMR of compound **12** with diastereotopic protons for CH$_2$ of C-4 phenylethyl group

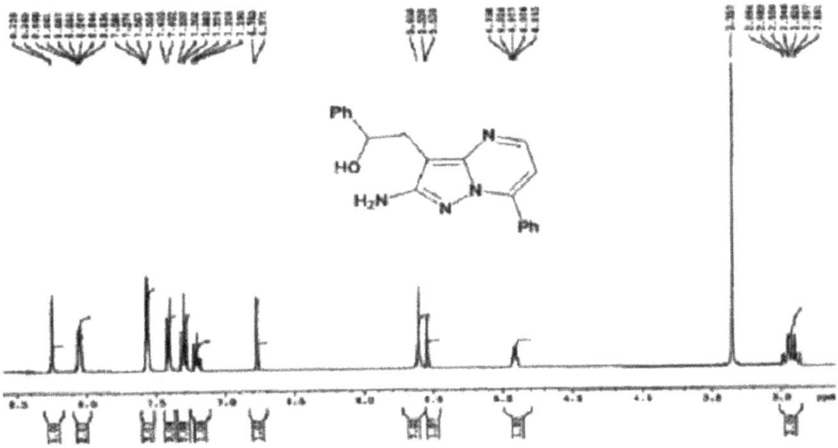

Fig. 27 Enlarged ¹HNMR of compound **12**

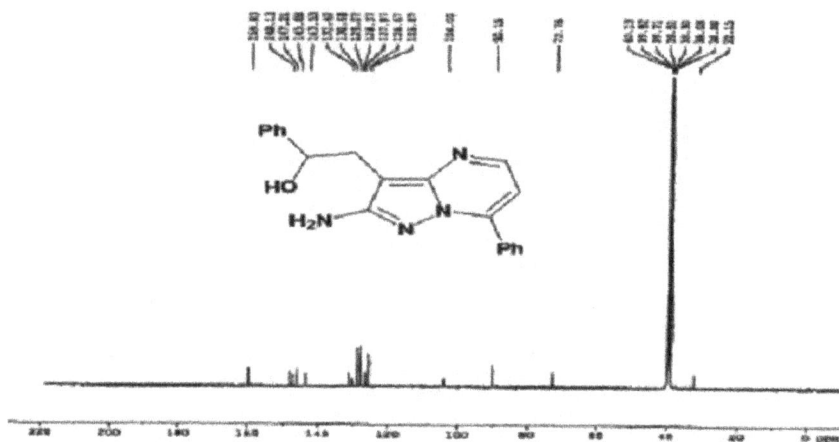

Fig. 28 ^{13}CNMR of compound **12**

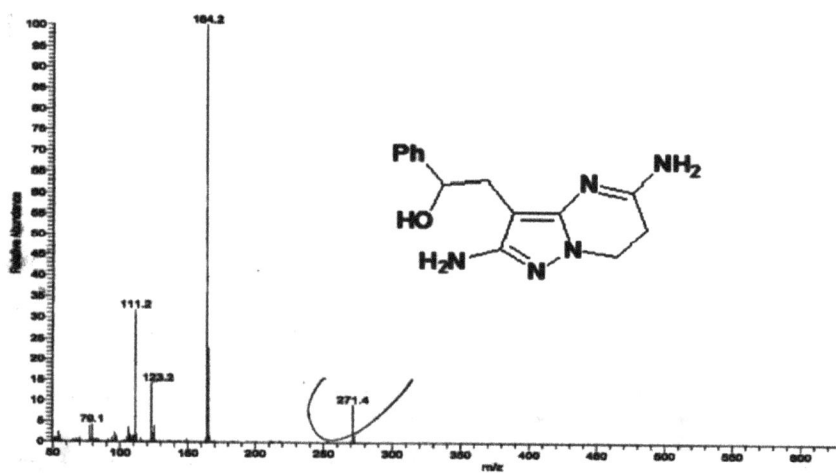

Fig. 29   Mass spectra of compound **13**

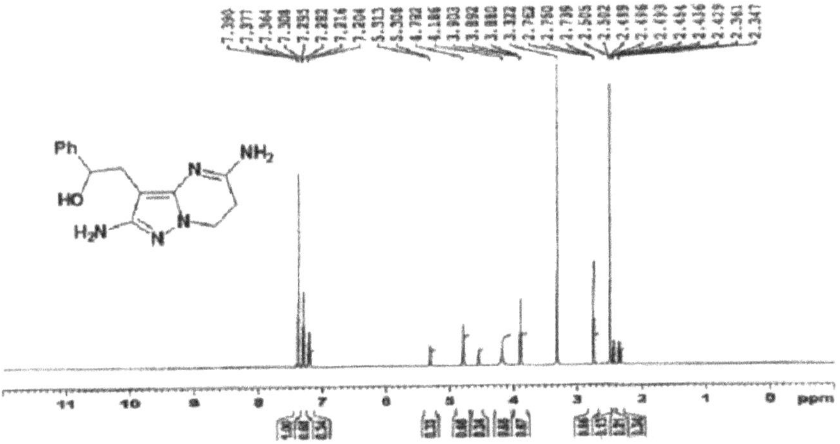

Fig. 30 ^{1}HNMR of comound **13** with the diastereotopic protons for CH$_2$ of the phenylethyl group at C-4

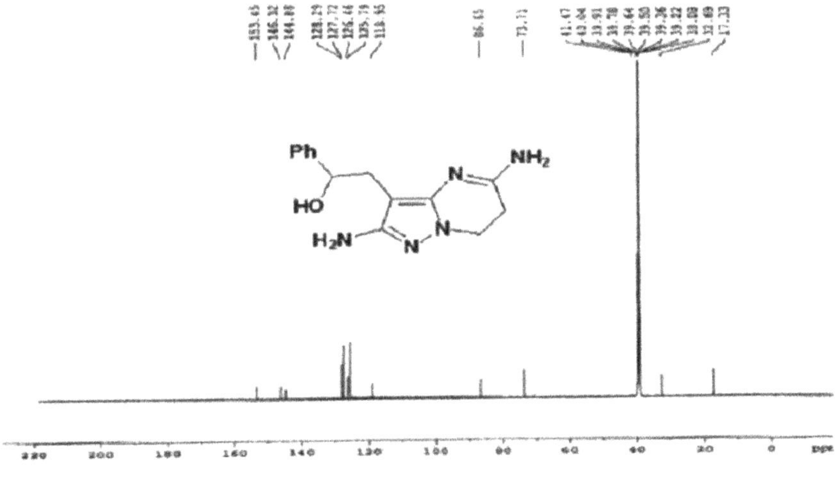

Fig. 31 ^{13}CNMR for compound **13**

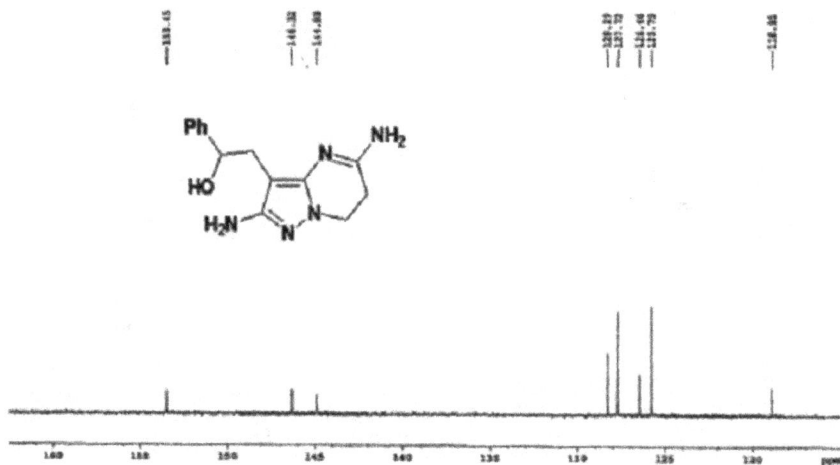

Fig. 32 ^{13}CNMR for compound **13**

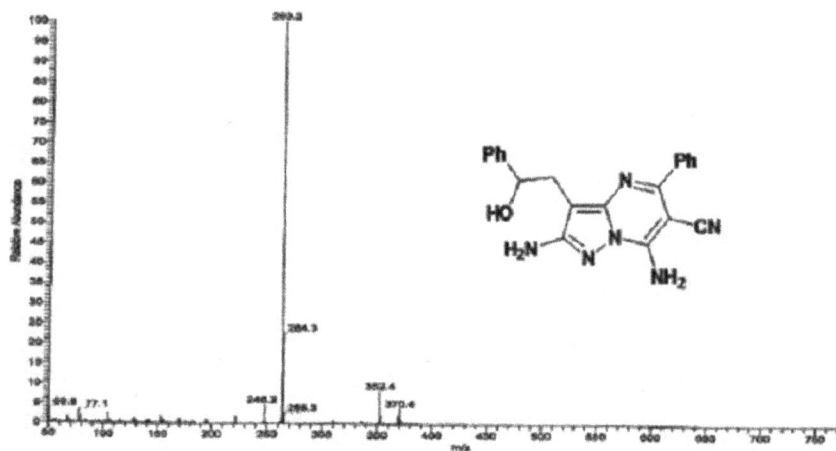

Fig. 33 Mass spectra of compound **13**

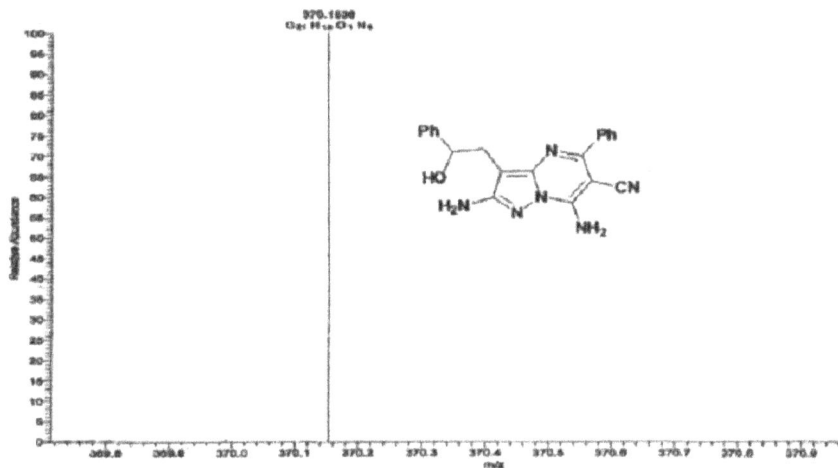

Fig. 34   Exact Mass of compound **14**

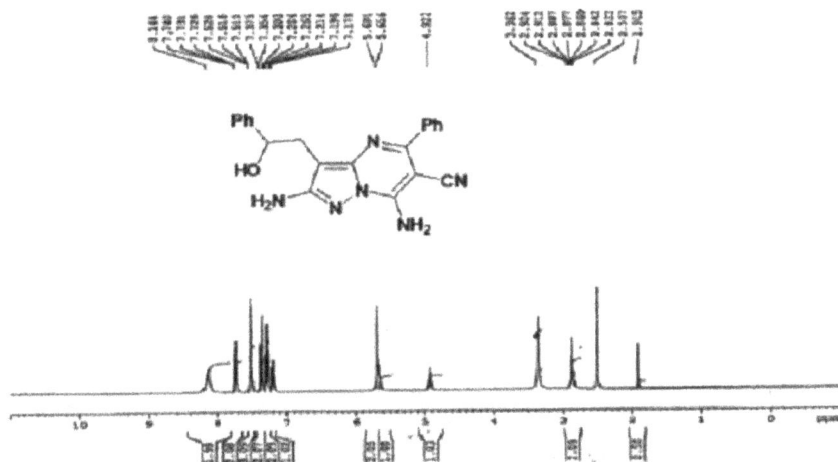

Fig. 35   ¹HNMR for compound **14** with diastereotopic protons for CH$_2$ of phenyle-thyl group at C-4

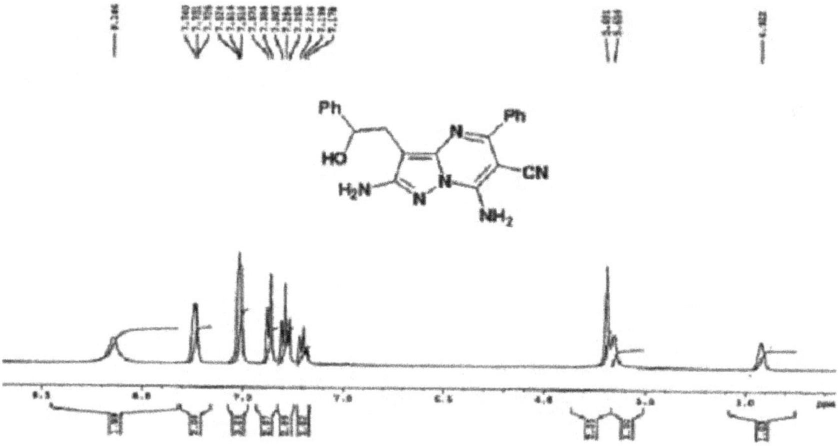

Fig. 36   Enlarged ^{1}HNMR of compound **14**

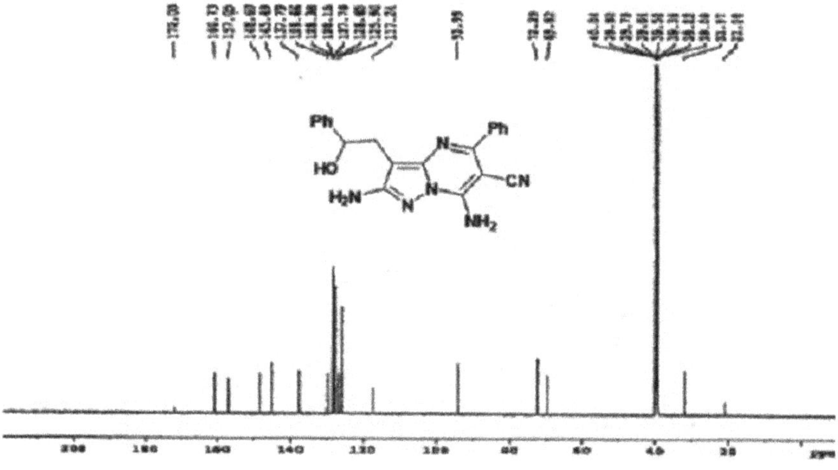

Fig. 37 ^{13}CNMR of compound **14**

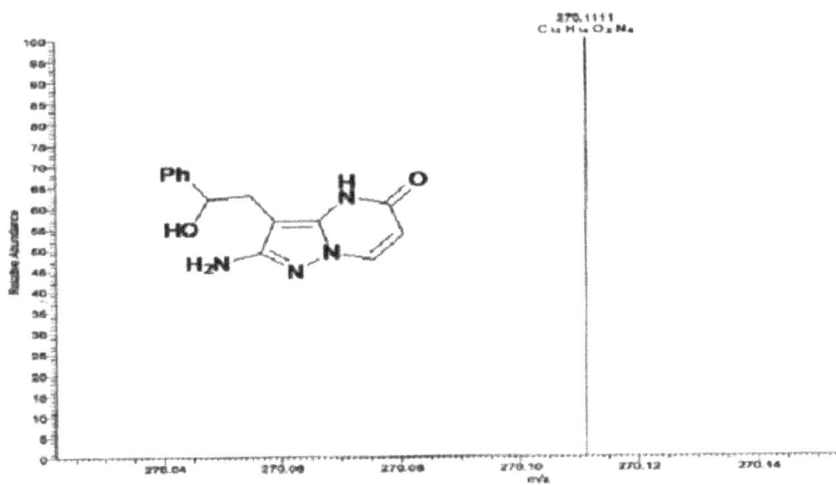

Fig. 38 Exact mass for the reaction product of **6** with diethyl malonate

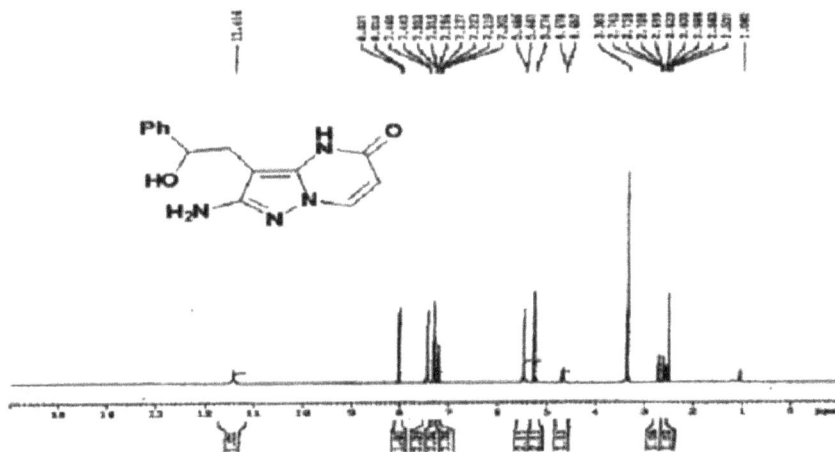

Fig. 39 ^{1}HNMR for the reaction product of **6** with diethyl malonate with diastereotopic protons for CH$_2$ of phenylethyl group at C-4

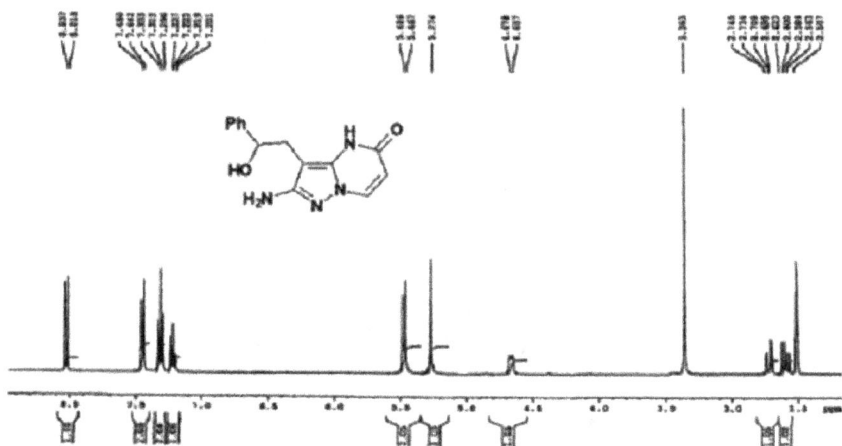

Fig. 40   Enlarged ^{1}HNMR for the reaction product of **6** with diethyl malonate

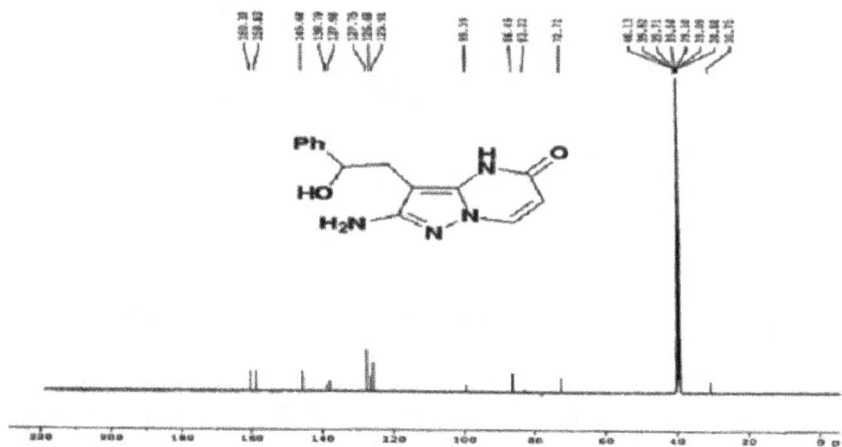

Fig. 41 ^{13}CNMR for the reaction of **6** with diethyl malonate

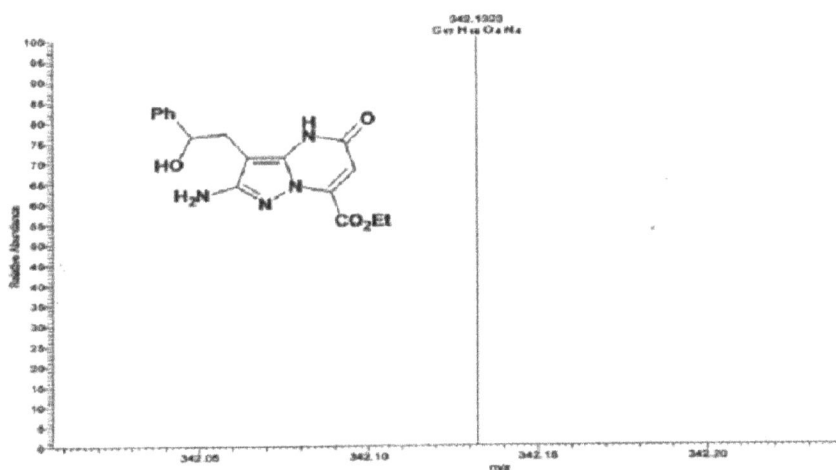

Fig. 42   Exact mass for the reaction product of **6** with diethyl acetylene dicarboxylate

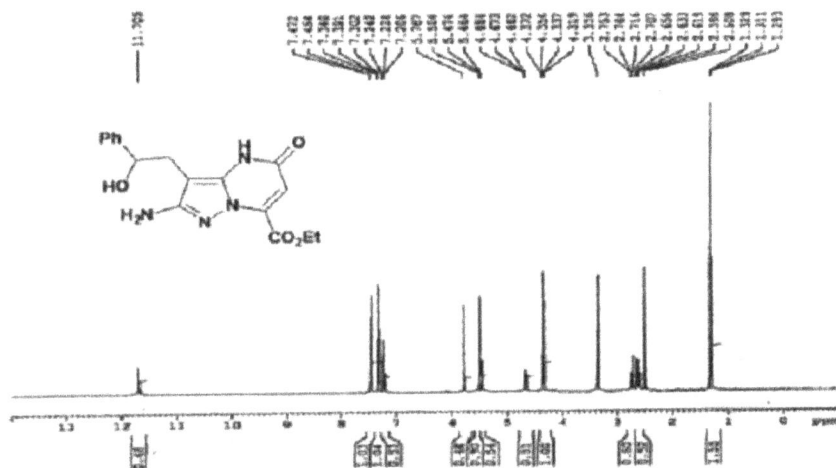

Fig. 43   ¹HNM for the reaction product of **6** with diethyl acetylene dicarboxylate with the diastereotropic protons for CH_2 of phenylethyl group at C-4

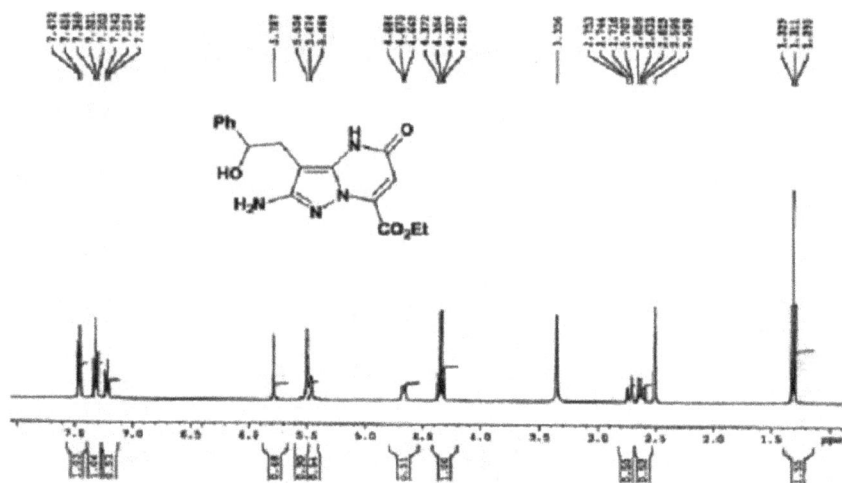

Fig. 44 Enlarged ^{1}HNMR for the reaction product of 6 with diethyl acetylene dicarboxylate

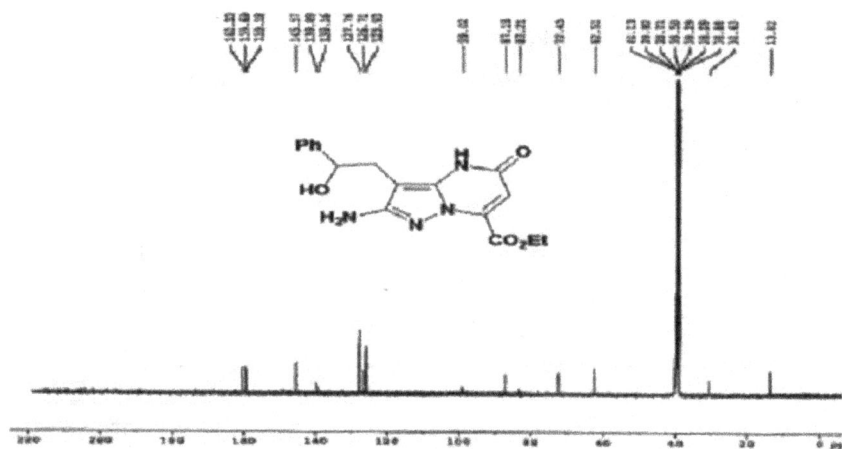

Fig. 45 ^{13}NMR for the reaction product of **6** with diethyl acetylene dicarboxylate

Reference

1. M. S. Moustafa, Ph.D. Thesis, South Valley University (**2014**).

5 Science Policy: Personal Experience

5.1. Introduction

It has been pointed out in The Economist that around 60% of research, especially in medicinal field, is intentionally distorted by avoiding reporting of side effects noted during research. In third-world countries, there is also plenty of misinformation. We will discuss its roots and include how, at least in Egypt, we can try to minimize this act.

I graduated from Cairo University 1962. I completed my M.Sc. and Ph.D. work between 1962 and 1969. All other information regarding my educational and professional background has been reported in several of my other books and review articles. Here, I will here emphasize the following:

1. How to select a research area
2. How to secure financial support
3. How to obtain background information on the subject
4. The aim of research in Universities
5. Ranking institutes and scientists
6. University regulations and their impact on research and University qualities
7. Experience of cooperation with:
   (a) Western institutes
   (b) Arab institutes
8. Quick summary of research achievements and also important conclusions.
9. Research ethics

177

5.2. How to Select a Research Area

As a new staff member, I observed that students selected the most productive staff member to ensure that they would obtain a degree within a reasonable amount of time. After almost sixty years in research, I have realized that this is almost always the method by which students select the type of research, and this is not going to change unless some rules are set by the institutions.

5.2.1. *How to secure financial support*

I went through different phases:

I- System finance is allocated to those who are well connected, not necessarily those who are productive; this is common to the whole Arab area, and even in some developed countries.

Initially, I received funding from the chemical industry in the west as my research was done during the golden days of new small molecules. Many multinational companies in my area were interested in their potential utilities in fields such as modern cosmetics, agrochemicals and as potential pharmaceuticals. Then, I explored further to make up for my deficiencies by training in the west. Again, those days were the golden days for the poor. I received funding for a year of training at the Tokyo Institute of Technology. In my days, the system was smart and left us to compete for funds in the international market. This is not possible now. The system is now wholly involved in this activity — for example, the well-known Alexander von Humboldt (AvH) fellows. Some of the funding I received back then were as follows:

1. Royal Norwegian council fellowship
2. AvH support (11 times)
3. British Council support
4. IOCD support, as no one helped me. I am also a free person and dictators do not like this. I am now focusing on how to attract students to this research area. In Cairo, only Christian researchers and poor ones were interested to cooperate as I do not care for the religion of student

and I finance through consultancy with multinationals working with all my students.
5.   Some students of the upper class also worked with me, and I think this was interesting type of work.

5.3. How can One Get the Background of the Subject?

Now, this is easy as databases are available, but I will give some examples here. I remember that plenty of my work was republished despite the fact that we had done the same several years earlier. Some of these articles has given our paper as the reference in introduction, and so when we raised a complaint, they stated that it was fine because they had cited our work.

This may be fine, but they did not clarify that we have carried out the same work before. Here are a few examples of some non-interpretable behaviors.

1.   In 1983, we reported the following[1]:

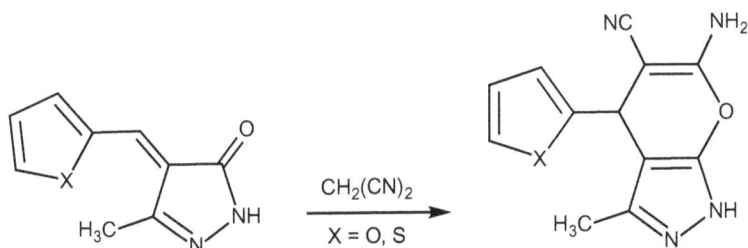

Five years later, totally ignoring any reference to this, Abdelrazek, Metz and Hanafy[2] reported the same. Is this science?

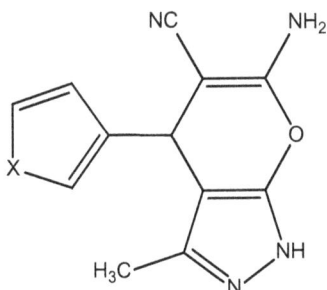

We then proved that the molecule in fact exists as

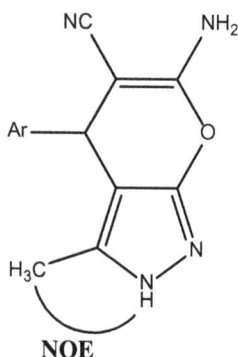

NOE

This has also been republished again, ignoring reference to our work. The referees in this case are also just as responsible. In the Ph.D. thesis of Y. R. Ibrahim (Minia University),[3] it was reported that the 3-oxoalkanonitriles readily undergo self-condensation to yield 2-aroylanilines on heating in pyridine for eight hours. Abdelrazek, in 2006, reported similar reaction with a different isolable product, which proved to be incorrect.[5]

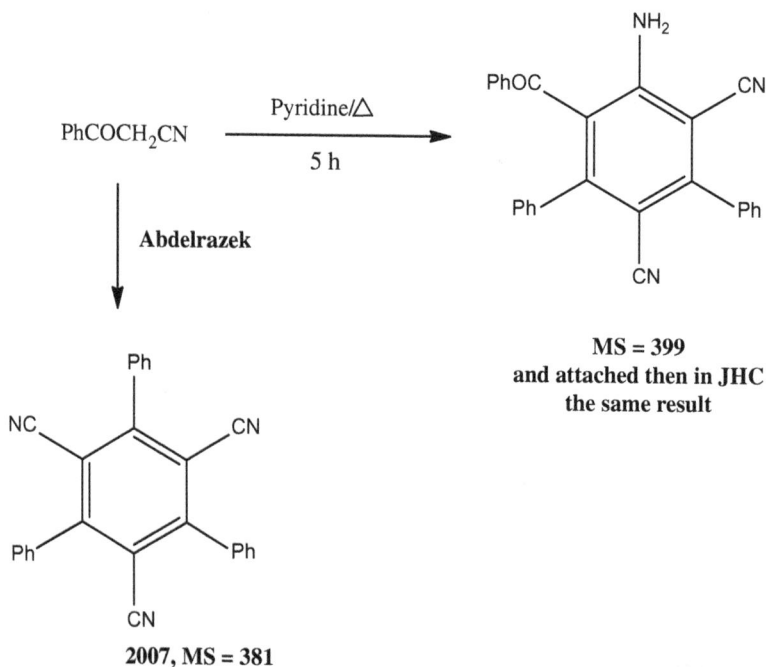

PhCOCH$_2$CN $\xrightarrow[\text{5 h}]{\text{Pyridine}/\triangle}$

Abdelrazek

MS = 399
and attached then in JHC
the same result

2007, MS = 381

This is not a discussion of who is correct, but in Abdelrazek paper, no reference to the Ph.D. thesis was given.

Should theses references be cited?

It has been initially reported that malononitrile undergoes 1,2-condensation with enaminone carbonyl function, and this was later corrected in *Molecules Journal*[6] (2009) and then with X-ray in *Arkivoc*.[7]

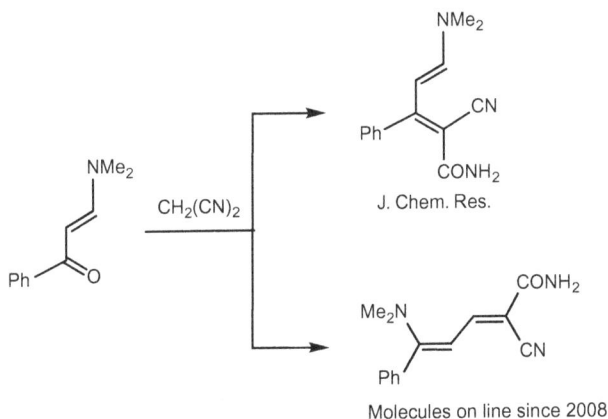

J. Chem. Res.

Molecules on line since 2008

Abdelrazek[8] claimed he has discovered that?. Is it correct for a close follower of my work in Researchgate to ignore that we have done this earlier?

In 1982, we reported on the chemistry of thiazole **1**; then in 1987, A. O. Abdelhamid reported on thiazole **2**, but the entire discussion looks to be type of plagiarism that is now prohibited.[9] Should recent references be cited in the discussions of paper be copied word for word from the 1982 paper?

Why does this happen only in Egyptian universities.

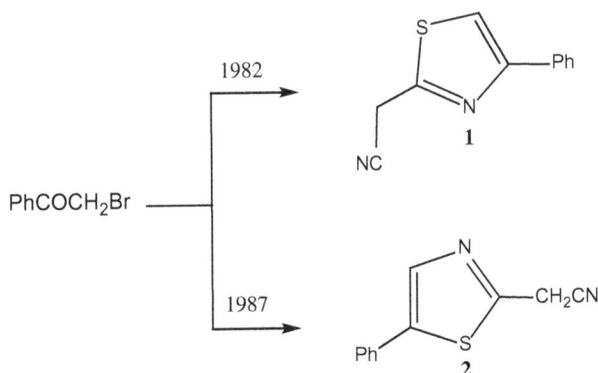

This is because of University regulations set in place before 1972. We used British regulations only. One Professor was appointed after 1972 regulations allowed self-breeding, and thus all worked following Elnagdi's or Shwally's directions. Also, some worked on the research area developed by Y. A. Ibrahim.

It was not a must to get the recommendation of the former supervisor, but attacking his work gives you privilege. Now, A. O. Abdelhamid is the Director of a promotion committee. This is because only the applicant's potential pilgrimages were looked at and no other aspects about him were considered.

F. M. Abdelrazik is also a member of promotion committees. In light of this, I decided that I could no longer accept being a member of such a corrupt place.

So, I shifted to Kuwait in 1993. But before I focus on yet another type of governmental corruption, I will try to give a key reason for the tendency to repeat the same data published by others or the same theory reported earlier using new substituents in the molecule and ignoring the existence of the originally produced work. It is in fact due to the lazy or ignorant referees since it is now easy to get access to a database that provides background regarding this work just by pressing a button on your phone.

Those who do this belong to two categories of researchers:

One tries to save cost of analysis while believing and accepting that almost everything that was done was correct and so depends on the Internet to get spectroscopic data. Perhaps now data from the Internet has become difficult to use as journals ask for originals as supplementary materials, but at least the researcher still saved time and practical experimentation by doing repetitive work.

The second group consists of those who claim to have arrived to new conclusions. They know that these are incorrect.

In this respect, I suggest the return of a magazine like organic synthesis where someone repeats experiments for at least those that are deemed to be a breakthrough. Universities should force staff seeking promotions to venture into new areas of research. Some complain of cost, but this is not true. Green methods should be more economic than classical methods.

If this is done, then with time, experts in every new field will accumulate in Egyptian Universities that are currently crowded with the so-called professors (most of these work only on heterocyclic chemistry).

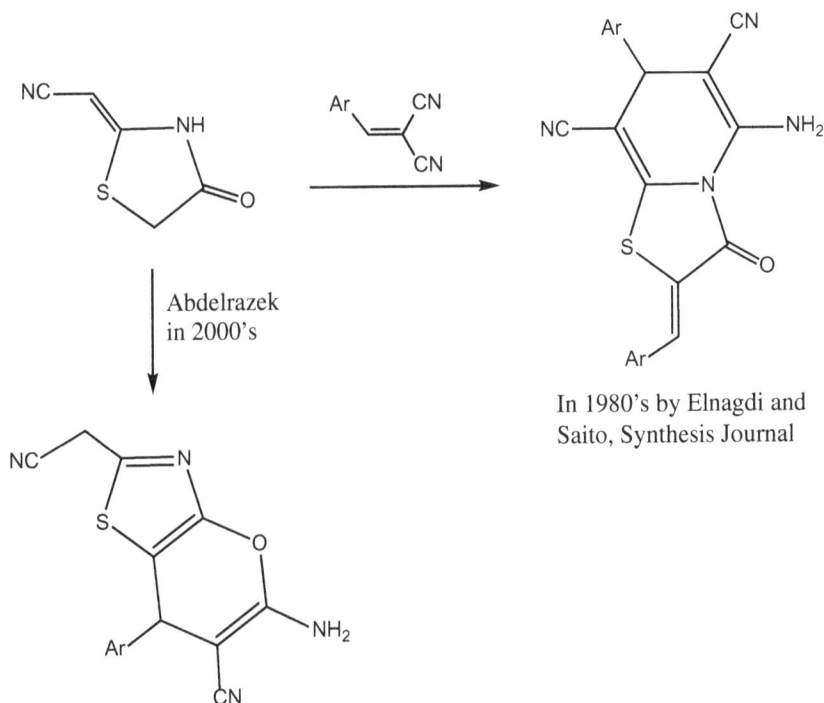

Abdelrazek
in 2000's

In 1980's by Elnagdi and
Saito, Synthesis Journal

Finally, without any comments, a copy of thesis was awarded in 2009, i.e. registered in 2006 and the author of the paper I can see is a member of the research committee and has signed that he attended the student presentation in 2006.

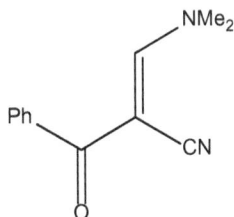

5.4. Conclusion

1. In Egyptian Universities, rules should be changed and supervisors' recommendation should be taken seriously in any further promotion.
2. Self-breeding should be stopped, giving chance for new fields to replace all fashioned ones.
3. Rules governing staff appointment should be changed. Vacancies should be announced even at the even international scale to provide an opportunity for the talented everywhere.
4. Universities should inspect the factuality of the data presented with those conducted at the analytical data units.

5.5. Summary of Confirmed Conformations and Their Significance

1. Synthesis of arylazo chalcones and their electrocyclization into cinnolines (Scheme 1).

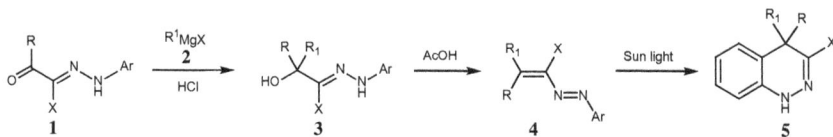

Scheme 1

2. Rearrangement of oxazolones and triazolones (patented latter as cannabinoid receptor inhibitors) (Scheme 2).

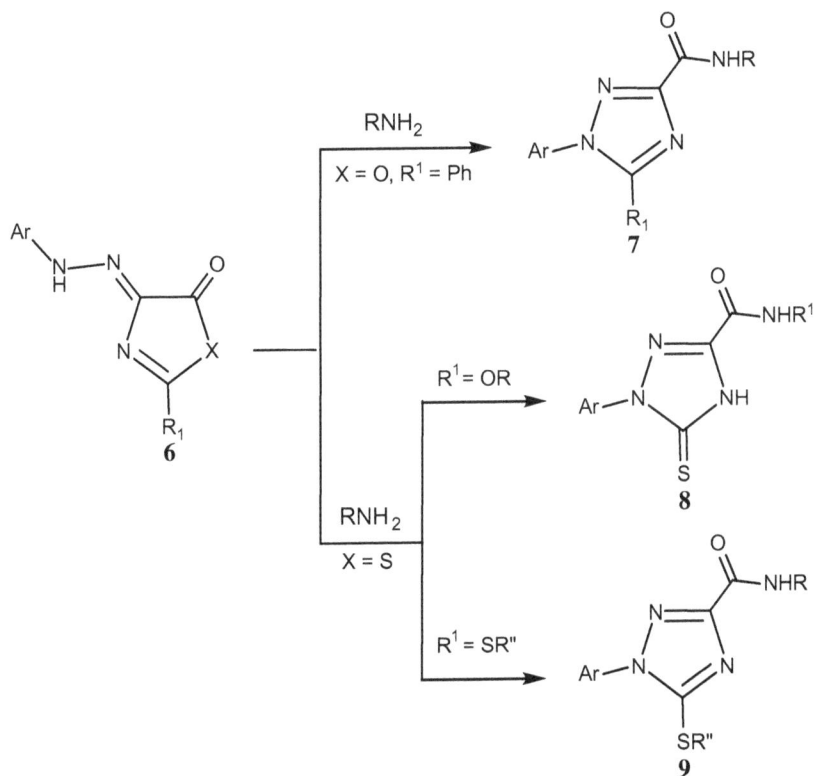

Scheme 2

3.   Chemistry of cinnamonitriles (many patented activities) (Scheme 3).

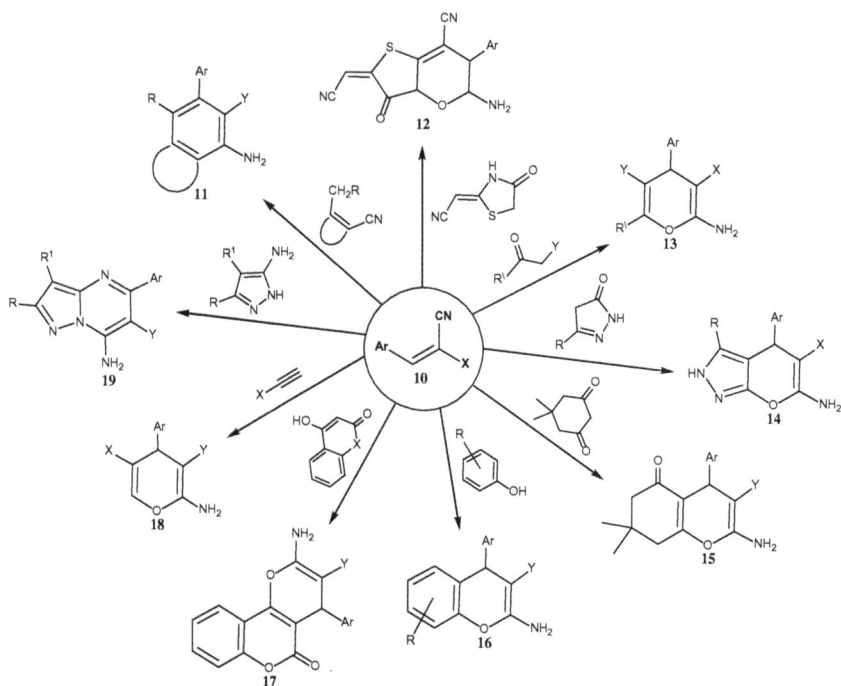

Scheme 3

In total, more than 1000 citations and several patented utilities.

4. New simple pyridazinone and condensed pyridazinone synthesis (Schemes 4 and 5).

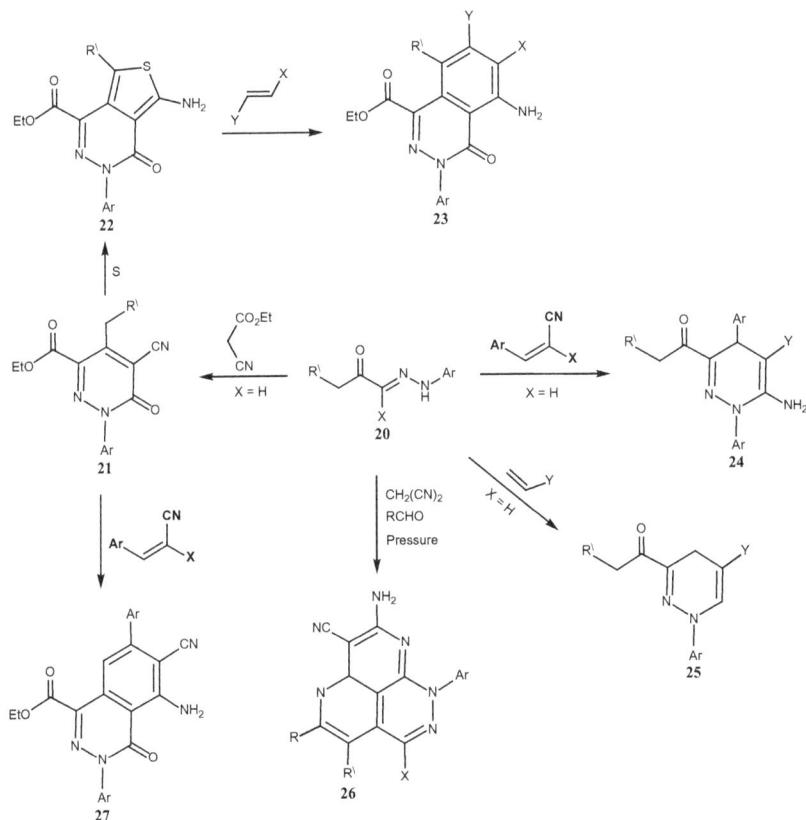

Scheme 4

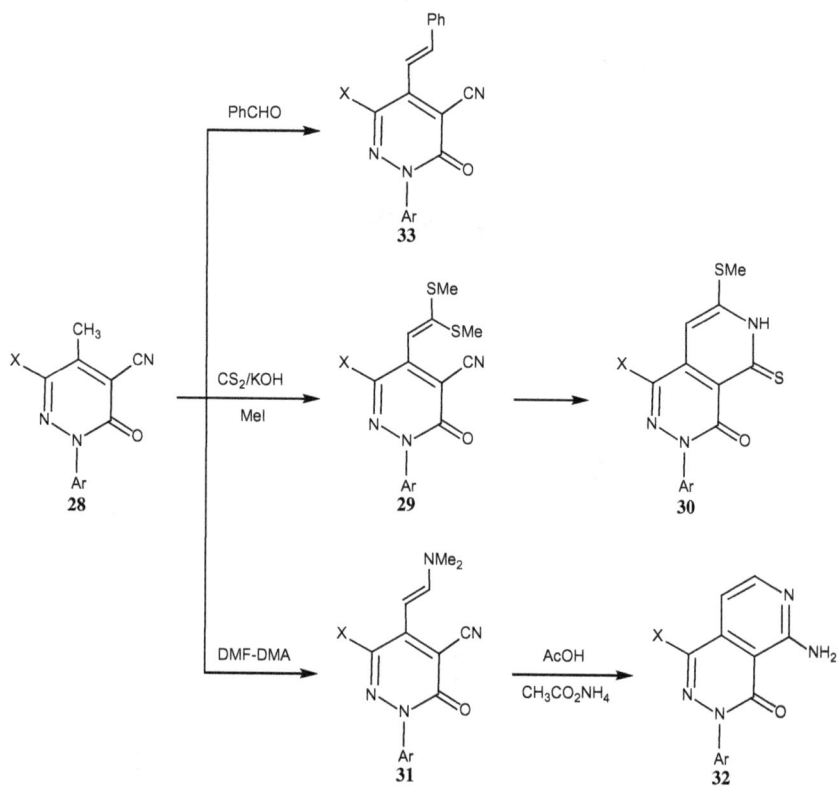

Scheme 5

5. Enamines (Scheme 6).
   (a) Elnagdi and Wamhoff as AvH fellow
   (b) Elnagdi

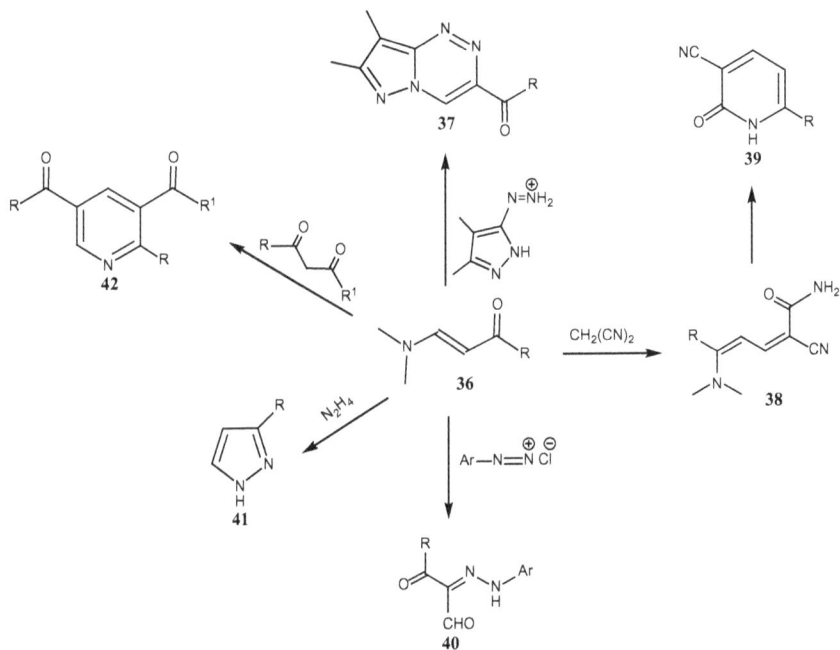

Scheme 6

6. Chemistry of arylhydrazonals (Scheme 7).

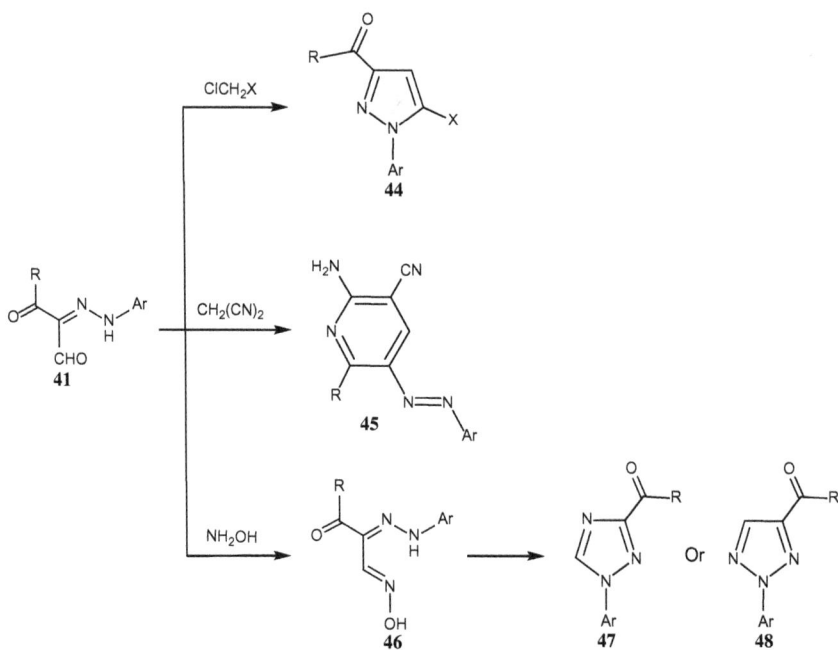

Scheme 7

7. Chemistry of 3-oxoalkanenitriles (Scheme 8).

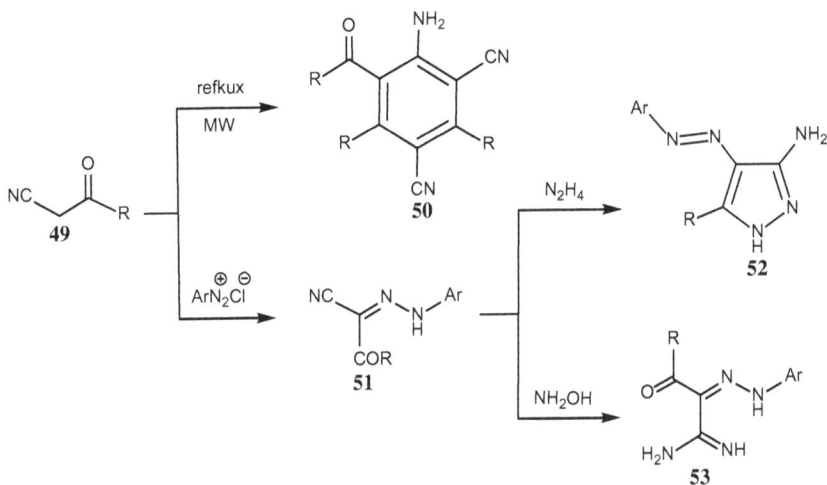

Scheme 8

References

1. N. S. Girgis, G. E. H. Elgemeie, G. A. M. Nawar and M. H. Elnagdi, *Liebigs Ann. Chem.*, *1468* (**1983**).
2. F. M. Abdelrazek, P. Metz, N. H. Metwally and S. F. El-Mahrouky, *Arch. Pharm. Chem. Life Sci. 339*, 456 (**2006**).
3. Y. S. Ibrahim, Ph.D. Thesis, Minia University (**1983**).
4. F. M. Abdelrazek and F. A. Michael, *J. Heterocycl. Chem. 43*, 7 (**2006**).
5. N. A. Al-Awadi, M. M. Abdelkhalik, I. A. Abdelhamid and M. H. Elnagdi, *Synlett*, *9*, 2979 (**2007**).
6. A. Alnajjar, M. M. Abdelkhalik, A. Al-Enzi and M. H. Elnagdi, *Molecules*, *14*, 68, (**2009**).
7. S. A. Al-Mousawi, M. S. Moustafa, M. M. Abdelkhalik and M. H. Elnagdi, *Arkivoc*, *xi*, 1 (**2009**).
8. F. M. Abdelrazik and A. N. Elsayed, *J. Heterocyclic Chem.*, *46*, 949 (**2009**).
9. M. H. Elnagdi, K. U. Sadek and M. S. Moustafa, *Adv. Heterocycl. Chem.*, *109*, 241 (**2013**).

.

Index